BIOZONE

Biology Modular Workbook Series

Evolution

The Biozone Writing Team:

Tracey Greenwood

Lyn Shepherd

Richard Allan

Daniel Butler

This first edition is dedicated to **Robert E. Seigman**, whose enthusiasm for the teaching of evolution continues to inspire

Published by:

Biozone International Ltd

109 Cambridge Road, Hamilton 3216, New Zealand

Printed by REPLIKA PRESS PVT LTD

Distribution Offices:

United Kingdom & Europe	**Biozone Learning Media (UK) Ltd**, Scotland
	Telephone: +44 (131) 557 5060
	Fax: +44 (131) 557 5030
	Email: sales@biozone.co.uk
	Website: www.biozone.co.uk
USA, Canada, South America, Africa	**Biozone International Ltd**, New Zealand
	Telephone: +64 (7) 856 8104
	Freefax: 1-800717-8751 (USA-Canada only)
	Fax: +64 (7) 856 9243
	Email: sales@biozone.co.nz
	Website: www.biozone.co.nz
Asia & Australia	**Biozone Learning Media Australia**, Australia
	Telephone: +61 (7) 5575 4615
	Fax: +61 (7) 5572 0161
	Email: sales@biozone.com.au
	Website: www.biozone.com.au

© 2006 Biozone International Ltd

First Edition 2006

ISBN: 1-877329-88-6

Front cover photographs:

Fossil Archaeopteryx.

3-D reconstruction of raptors, created by Dan Butler using Poser IV, Curious Labs.

Biology Modular Workbook Series

The Biozone *Biology Modular Workbook Series* has been developed to meet the demands of customers with the requirement for a modular resource which can be used in a flexible way. Like Biozone's popular Student Resource and Activity Manuals, these workbooks provide a collection of visually interesting and accessible activities, which cater for students with a wide range of abilities and background. The workbooks are divided into a series of chapters, each comprising an introductory section with detailed learning objectives and useful resources, and a series of write-on activities ranging from paper practicals and data handling exercises, to questions requiring short essay style answers. Material for these workbooks has been drawn from Biozone's popular, widely used manuals, but the workbooks have been structured with greater ease of use and flexibility in mind. During the development of this series, we have taken the opportunity to improve the design and content, while retaining the basic philosophy of a student-friendly resource which spans the gulf between textbook and study guide. With its unique, highly visual presentation, it is possible to engage and challenge students, increase their motivation and empower them to take control of their learning.

Evolution

This title in the *Biology Modular Workbook Series* provides students with a set of comprehensive guidelines and highly visual worksheets through which to explore aspects of population genetics and evolution. *Evolution* is the ideal companion for students of the life sciences, encompassing the origins of life on Earth, the evidence for evolution, gene pools and microevolution, and large scale evolutionary patterns such as adaptive radiation and extinction. This workbook comprises three chapters, each covering a different aspect of evolutionary biology. These areas are explained through a series of activities, each of which explores a specific concept (e.g. speciation or coevolution). Model answers (on CD-ROM) accompany each order free of charge. *Evolution* is a student-centred resource. Students completing the activities, in concert with their other classroom and practical work, will consolidate existing knowledge and develop and practise skills that they will use throughout their course. This workbook may be used in the classroom or at home as a supplement to a standard textbook. Some activities are introductory in nature, while others may be used to consolidate and test concepts already covered by other means. Biozone has a commitment to produce a cost-effective, high quality resource, which acts as a student's companion throughout their biology study. Please do not photocopy from this workbook; we cannot afford to provide single copies of workbooks to schools and continue to develop, update, and improve the material they contain.

Acknowledgements and Photo Credits

Biozone's authors also acknowledge the generosity of those who have kindly provided information or photographs for this edition (some identified by way of coded credits): • Dept of Conservation for their invaluable assistance, especially Ferne McKenzie, for access to the DoC photo library • **Sean Carroll** and his text "*Endless forms most beautiful*" for authoritative information and examples for the new material on evolutionary developmental biology • Dr. Alan Cooper, Smithsonian Institute, for information on ratite evolution • The late Ron Lind, for photographs of stromatolites • Dr. John Green for his input to the topics on evolution • Dr. John Stencel for his data on the albino grey squirrel population • Ewan Grant-Mackie and Prof. J.M. Grant-Mackie for information on the adaptive radiation of NZ wrens and the evolution of NZ parrots • Assoc. Prof. Bruce Clarkson, University of Waikato/CBER, for his contributions towards the activities on adaptive radiation in *Hebe* • Dept. of Natural Resources, Illinois, for information on genetic diversity in prairie chickens • Liam Nolan for his contributions to the activities on the genetic biodiversity of Antarctic springtails • Missouri Botanical Gardens for their photograph of egg mimicry in *Passiflora* • Alex Wild for his photograph of swollen thorn *Acacia* • California Academy of Sciences for the photograph of a Galapagos ground finch • Leo Sanchez and Burkhard Budel for use of their photographs used in the activities on Antarctic springtails. Royalty free images, purchased by Biozone International Ltd, are used throughout this manual and have been obtained from the following sources: istockphotos (www.istockphoto.com) • Corel Corporation from various titles in their Professional Photos CD-ROM collection; ©Hemera Technologies Inc, 1997-2001; © 2005 JupiterImages Corporation www.clipart.com; PhotoDisc®, Inc. USA, www.photodisc.com. Coded credits as follows: **DoC**: Department of Conservation (NZ), **DoC-CV**: C.R. Veitch, **DoC-RM**: Rod Morris, **EII**: Education Interactive Imaging, **NASA**: National Aeronautics and Space Administration, **RA**: Richard Allan, **RCN**: Ralph Cocklin, **RL**: Ron Lind.

Also in this series:

Skills in Biology

Cell Biology & Biochemistry

Genes & Inheritance

Human Evolution

For other titles in this series go to: www.thebiozone.com/modular.html

Contents

Activity is marked: • to be done; ✔ when completed

How to Use this Workbook

Evolution is designed to provide students with a resource that will make the acquisition of knowledge and skills in this area easier and more enjoyable. An appreciation of the overwhelming evidence for evolution, and an understanding of the principles of evolutionary biology are important in most biology curricula. Moreover, this subject is of high interest; it is both relevant and, for some, controversial. What is more, the application of new gene technologies is increasingly adding to our understanding of evolutionary mechanisms. This workbook is suitable for all students of the life sciences, and will reinforce and extend the ideas developed by teachers. It is **not a textbook**; its aim is to complement the texts written for your particular course. Each chapter in *Evolution* provides the following resources. You should refer back to them as you work through each set of worksheets.

Guidance Provided for Each Topic

Learning objectives:

These provide you with a map of the chapter content. Completing the learning objectives relevant to your course will help you to satisfy the knowledge requirements of your syllabus. Your teacher may decide to leave out points or add to this list.

Chapter content:

The upper panel of the header identifies the general content of the chapter. The lower panel provides a brief summary of the chapter content.

Key words:

Key words are displayed in **bold** type in the learning objectives and should be used to create a glossary as you study each topic. From your teacher's descriptions and your own reading, write your own definition for each word.

Note: Only the terms relevant to your selected learning objectives should be used to create your glossary. Free glossary worksheets are also available from our web site.

Use the check boxes to mark objectives to be completed.
Use a **dot** to be done (•).
Use a **tick** when completed (✓).

Supplementary texts:

References to supplementary texts suitable for use with this workbook are provided. The details of these are provided on page 7, together with other resources information.

Periodical articles:

Ideal for those seeking more depth or the latest research on a specific topic. Articles are sorted according to their suitability for student or teacher reference. Visit your school, public, or university library for these articles.

Internet addresses:

Access our database of links to more than **800** web sites (updated regularly) relevant to the topics covered. Go to Biozone's own web site: **www.thebiozone.com** and link directly to listed sites using the *BioLinks* button.

Supplementary resources

Biozone's Presentation MEDIA are noted where appropriate.

Activity Pages

The activities and exercises make up most of the content of this workbook. They are designed to reinforce the concepts you have learned about in the topic. Your teacher may use the activity pages to introduce a topic for the first time, or you may use them to revise ideas already covered. They are excellent for use in the classroom, and as homework exercises and revision. In most cases, the activities should not be attempted until you have carried out the necessary background reading from your textbook. As a self-check, model answers for each activity are provided on CD-ROM with each order of workbooks.

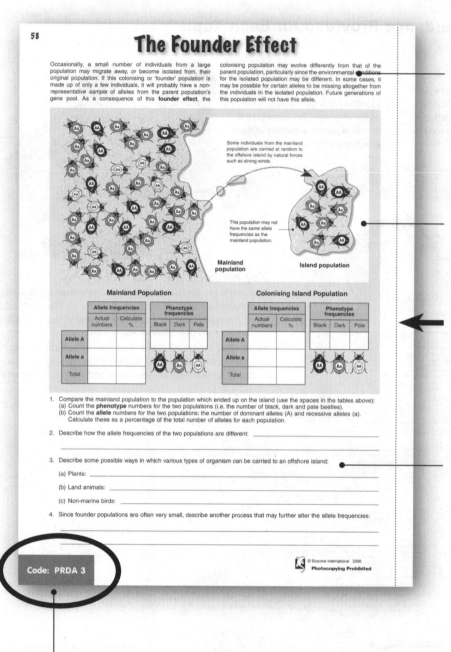

Introductory paragraph:
The introductory paragraph sets the 'scene' for the focus of the page and provides important background information. Note any words appearing in **bold**; these are 'key words' which could be included in a glossary of biological terms for the topic.

Easy to understand diagrams:
The main ideas of the topic are represented and explained by clear, informative diagrams.

Tear-out pages:
Each page of the book has a perforation that allows easy removal. Your teacher may ask you to remove activity pages for marking, or so that they can be placed in a ringbinder with other work on the topic.

Write-on format:
You can test your understanding of the main ideas of the topic by answering the questions in the spaces provided. Where indicated, your answers should be concise. Questions requiring explanation or discussion are spaced accordingly. Answer the questions appropriately according to the specific questioning term used (see the facing page).

Activity code:
Activities are coded to help you in identifying the type of activities and the skills they require. Most activities require some basic knowledge recall, but will usually build on this to include applying the knowledge to explain observations or predict outcomes. The least difficult questions generally occur early in the activity, with more challenging questions towards the end of the activity.

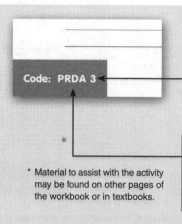

* Material to assist with the activity may be found on other pages of the workbook or in textbooks.

Activity Level

1 = Simple questions not requiring complex reasoning
2 = Some complex reasoning may be required
3 = More challenging, requiring integration of concepts

Type of Activity

D = Includes some data handling and/or interpretation
P = includes a paper practical
R = May require research outside the information on the page, depending on your knowledge base*
A = Includes application of knowledge to solve a problem
E = Extension material

Explanation of Terms

Questions come in a variety of forms. Whether you are studying for an exam or writing an essay, it is important to understand exactly what the question is asking. A question has two parts to it: one part of the question will provide you with information, the second part of the question will provide you with instructions as to how to answer the question. Following these instructions is most important. Often students in examinations know the material but fail to follow instructions and do not answer the question appropriately. Examiners often use certain key words to introduce questions. Look out for them and be clear as to what they mean. Below is a description of terms commonly used when asking questions in biology.

Commonly used Terms in Biology

The following terms are frequently used when asking questions in examinations and assessments. Students should have a clear understanding of each of the following terms and use this understanding to answer questions appropriately.

Account for: Provide a satisfactory explanation or reason for an observation.

Analyse: Interpret data to reach stated conclusions.

Annotate: Add **brief** notes to a diagram, drawing or graph.

Apply: Use an idea, equation, principle, theory, or law in a new situation.

Appreciate: To understand the meaning or relevance of a particular situation.

Calculate: Find an answer using mathematical methods. Show the working unless instructed not to.

Compare: Give an account of similarities and differences between two or more items, referring to both (or all) of them throughout. Comparisons can be given using a table. Comparisons generally ask for similarities more than differences (see contrast).

Construct: Represent or develop in graphical form.

Contrast: Show differences. Set in opposition.

Deduce: Reach a conclusion from information given.

Define: Give the precise meaning of a word or phrase as concisely as possible.

Derive: Manipulate a mathematical equation to give a new equation or result.

Describe: Give a detailed account, including all the relevant information.

Design: Produce a plan, object, simulation or model.

Determine: Find the only possible answer.

Discuss: Give an account including, where possible, a range of arguments, assessments of the relative importance of various factors, or comparison of alternative hypotheses.

Distinguish: Give the difference(s) between two or more different items.

Draw: Represent by means of pencil lines. Add labels unless told not to do so.

Estimate: Find an approximate value for an unknown quantity, based on the information provided and application of scientific knowledge.

Evaluate: Assess the implications and limitations.

Explain: Give a clear account including causes, reasons, or mechanisms.

Identify: Find an answer from a number of possibilities.

Illustrate: Give concrete examples. Explain clearly by using comparisons or examples.

Interpret: Comment upon, give examples, describe relationships. Describe, then evaluate.

List: Give a sequence of names or other brief answers with no elaboration. Each one should be clearly distinguishable from the others.

Measure: Find a value for a quantity.

Outline: Give a brief account or summary. Include essential information only.

Predict: Give an expected result.

Solve: Obtain an answer using algebraic and/or numerical methods.

State: Give a specific name, value, or other answer. No supporting argument or calculation is necessary.

Suggest: Propose a hypothesis or other possible explanation.

Summarise: Give a brief, condensed account. Include conclusions and avoid unnecessary details.

In Conclusion

Students should familiarise themselves with this list of terms and, where necessary throughout the course, they should refer back to them when answering questions. The list of terms mentioned above is not exhaustive and students should compare this list with past examination papers / essays etc. and add any new terms (and their meaning) to the list above. The aim is to become familiar with interpreting the question and answering it appropriately.

Using the Internet

The internet is a vast global network of computers connected by a system that allows information to be passed through telephone connections. When people talk about the internet they usually mean the **World Wide Web** (WWW). The WWW is a service that has made the internet so simple to use that virtually anyone can find their way around, exchange messages, search libraries and perform all manner of tasks. The internet is a powerful resource for locating information. Listed below are two journal articles worth reading. They contain useful information on what the internet is, how to get started, examples of useful web sites, and how to search the internet.

- **Click Here: Biology on the Internet** Biol. Sci. Rev., 10(2) November 1997, pp. 26-29.
- **An A-level biologists guide to The World Wide Web** Biol. Sci. Rev., 10(4) March 1998, pp. 26-29.

Using the Biozone Website: www.thebiozone.com

The **Back** and **Forward** buttons allow you to navigate between pages displayed on a www site

The current **internet address (URL)** for the web site is displayed here. You can type in a new address directly into this space.

Tool bar provides a row of buttons with shortcuts for some commonly performed tasks, such as search functions or bookmarking a page.

Biozone International: biology resources.

http://www.thebiozone.com   Google

BIOZONE

Home | Products | Purchase Online | Biolinks | Resources | Free Samples | News | Contact

Biology Workbooks

Engaging informative activities and comprehensive learning objectives for selected topics in biology and complete biology programs

Contact a Biozone Office

Presentation Media

Bring your presentations to life ...

Presentation Media

A powerful and highly engaging series of presentation titles to enhance your lectures. Provided in multiple formats:

- PowerPoint
- Keynote
- Acrobat PDF
- QuickTime slideshow.

Includes a generous site licence.

Resources for teachers and students   Science Supplies, Biology Software and Videos

Biozone News

Up to date information on products, new releases and conference workshops.

Biolinks

Biozone's extensive database of Biolinks provides FREE access to:

- selected websites
- web based resources
- RSS newsfeeds.

Free Samples

FREE samples of many of Biozone's products are provided, including hundreds of previews of pages and screenshots.

Searching the Net

The WWW addresses listed throughout the manual have been selected for their relevance to the topic in which they are listed. We believe they are good sites. Don't just rely on the sites that we have listed. Use the powerful 'search engines', which can scan the millions of sites for useful information. Here are some good ones to try:

Alta Vista:	**www.altavista.com**
Ask Jeeves:	**www.ask.com**
Excite:	**www.excite.com/search**
Google:	**www.google.com**
Go.com:	**www.go.com**
Lycos:	**www.lycos.com**
Metacrawler:	**www.metacrawler.com**
Yahoo:	**www.yahoo.com**

Biozone International provides a service on its web site that links to all internet sites listed in this workbook. Our web site also provides regular updates with new sites listed as they come to our notice and defunct sites deleted. Our **BIO LINKS** page, shown below, will take you to a database of regularly updated links to more than 800 other quality biology web sites.

The **Resource Hub**, accessed via the homepage or resources, provides links to the supporting resources referenced in the workbook. These resources include comprehensive and supplementary texts, biology dictionaries, computer software, videos, and science supplies. These can be used to enhance your learning experience.

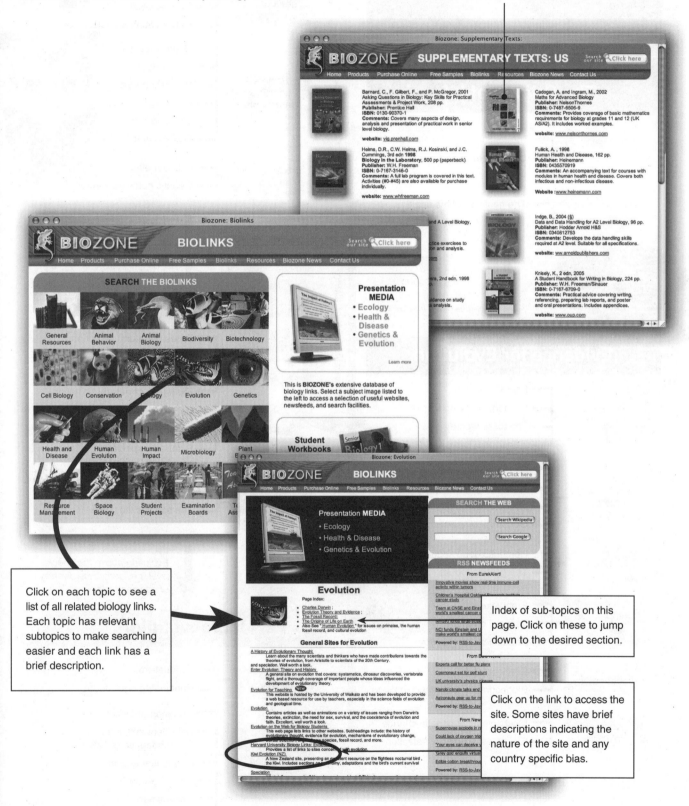

Click on each topic to see a list of all related biology links. Each topic has relevant subtopics to make searching easier and each link has a brief description.

Index of sub-topics on this page. Click on these to jump down to the desired section.

Click on the link to access the site. Some sites have brief descriptions indicating the nature of the site and any country specific bias.

Concept Map for Evolution

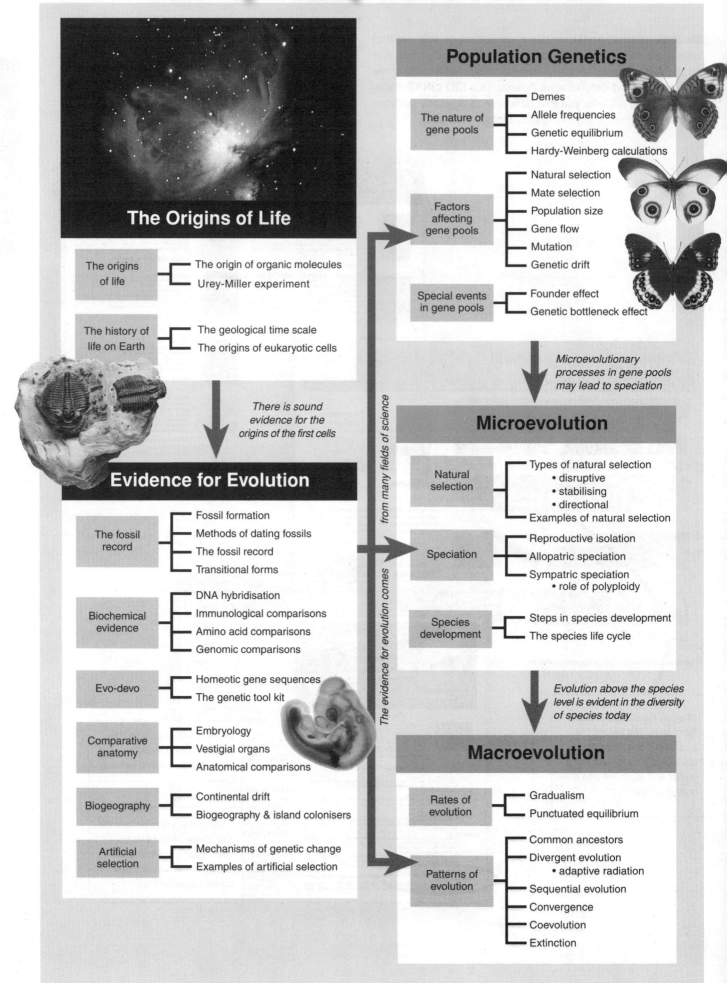

The Origins of Life

| The origins of life | The origin of organic molecules |
| | Urey-Miller experiment |

| The history of life on Earth | The geological time scale |
| | The origins of eukaryotic cells |

There is sound evidence for the origins of the first cells

Evidence for Evolution

The fossil record	Fossil formation
	Methods of dating fossils
	The fossil record
	Transitional forms

Biochemical evidence	DNA hybridisation
	Immunological comparisons
	Amino acid comparisons
	Genomic comparisons

| Evo-devo | Homeotic gene sequences |
| | The genetic tool kit |

Comparative anatomy	Embryology
	Vestigial organs
	Anatomical comparisons

| Biogeography | Continental drift |
| | Biogeography & island colonisers |

| Artificial selection | Mechanisms of genetic change |
| | Examples of artificial selection |

from many fields of science

The evidence for evolution comes

Population Genetics

The nature of gene pools	Demes
	Allele frequencies
	Genetic equilibrium
	Hardy-Weinberg calculations

Factors affecting gene pools	Natural selection
	Mate selection
	Population size
	Gene flow
	Mutation
	Genetic drift

| Special events in gene pools | Founder effect |
| | Genetic bottleneck effect |

Microevolutionary processes in gene pools may lead to speciation

Microevolution

Natural selection	Types of natural selection
	• disruptive
	• stabilising
	• directional
	Examples of natural selection

Speciation	Reproductive isolation
	Allopatric speciation
	Sympatric speciation
	• role of polyploidy

| Species development | Steps in species development |
| | The species life cycle |

Evolution above the species level is evident in the diversity of species today

Macroevolution

| Rates of evolution | Gradualism |
| | Punctuated equilibrium |

Patterns of evolution	Common ancestors
	Divergent evolution
	• adaptive radiation
	Sequential evolution
	Convergence
	Coevolution
	Extinction

Resources Information

Your set textbook should always be a starting point for information, but there are also many other resources available. A list of readily available resources is provided below. Access to the publishers of these resources can be made directly from Biozone's web site through our resources hub: **www.thebiozone.com/resource-hub.html**. Please note that our listing of any product in this workbook does not denote Biozone's endorsement of it.

Biology Dictionaries

Access to a good biology dictionary is useful when dealing with biological terms. Some of the titles available are listed below. Link to the relevant publisher via Biozone's resources hub or by typing: **www.thebiozone.com/resources/dictionaries-pg1.html**

Clamp, A. **AS/A-Level Biology. Essential Word Dictionary**, 2000, 161 pp. Philip Allan Updates. **ISBN**: 0-86003-372-4.
Carefully selected essential words for AS and A2. Concise definitions are supported by further explanation and illustrations where required.

Hale, W.G., J.P. Margham, & V.A. Saunders. **Collins: Dictionary of Biology** 3 ed. 2003, 672 pp. HarperCollins. **ISBN**: 0-00-714709-0.
Updated to take in the latest developments in biology from the Human Genome Project to advancements in cloning (new edition pending).

Henderson, I.F, W.D. Henderson, and E. Lawrence. **Henderson's Dictionary of Biological Terms**, 1999, 736 pp. Prentice Hall. **ISBN**: 0582414989
This edition has been updated, rewritten for clarity, and reorganised for ease of use. An essential reference and the dictionary of choice for many.

Lincoln, R.J., G.A. Boxshall, & P.F. Clark. **A Dictionary of Ecology, Evolution, and Systematics**, 2 ed., 1998, 371 pp. Cambridge Uni. Press. **ISBN**: 052143842X
6500 entries covering all major fields within biology, and expanded to reflect recent developments in the science. No pronunciation guidelines are provided.

McGraw-Hill (ed). **McGraw-Hill Dictionary of Bioscience**, 2 ed., 2002, 662 pp. McGraw-Hill. **ISBN**: 0-07-141043-0
22 000 entries encompassing more than 20 areas of the life sciences. It includes synonyms, acronyms, abbreviations, and pronunciations for all terms.

Supplementary Texts

Clegg, C.J., 1999. **Genetics and Evolution**, 96 pp. **ISBN**: 0-7195-7552-4
Concise but thorough coverage of molecular genetics, genetic engineering, inheritance, and evolution. An historical perspective is included by way of introduction, and a glossary and a list of abbreviations used are included.

Futuyma, D.J., 2005 **Evolution**, 543 pp (hardback) **Publisher**: Sinauer Associates **ISBN**: 0-8789-3187-2
Comments:A comprehensive text suitable as a teacher or student reference. Covers neo-Darwinism as well as excellent treatment of newer areas: the evolution of genes and genomes and development and evolution.

Helms, D.R., C.W. Helms, R.J. Kosinski, and J.C. Cummings, 3rd edn 1998 **Biology in the Laboratory**, 500 pp (softback) **Publisher**: W.H. Freeman **ISBN**: 0-7167-3146-0
Comments: *A full lab program is covered in this text. Activities (#0-#45) are also available for purchase individually.*

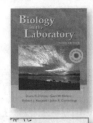

Jones, N., A. Karp., & G. Giddings, 2001. **Essentials of Genetics**, 224 pp. **ISBN**: 0-7195-8611-9
A thorough and very readable supplement for genetics and evolution. Comprehensive coverage of cell division, molecular genetics, and genetic engineering. The application of new gene technologies to agriculture, industry, and medicine is also discussed.

Martin, R.A., 2004. **Missing Links**, 302 pp. **ISBN**: 0-7637-2196-4
This book introduces evolutionary science with an accessible discussion of basic scientific practices, rock and fossil dating techniques, and schools of classification. Examples of evolutionary transition are provided, from the origins of life to the morphological changes that readers will observe in their lifetimes.

Zimmer, C., 2001 **Evolution: The Trumph of an Idea**, 362 pp. (hardback) **Publisher**: HarperCollins **ISBN**: 0-06-019906-7
Comments: *A vivid, accessible exploration of Darwin's ideas; essential reading for those who wish to know how life arose and diversified.*

Periodicals, Magazines, & Journals

Biological Sciences Review: *An informative quarterly publication for biology students.* Enquiries: **UK**: Philip Allan Publishers **Tel**: 01869 338652 **Fax**: 01869 338803 **E-mail**: sales @philipallan.co.uk **Australasia**: **Tel**: 08 8278 5916, **E-mail**: rjmorton@ adelaide.on.net

New Scientist: *Widely available weekly magazine with research summaries and features.* Enquiries: Reed Business Information Ltd, 51 Wardour St. London WIV 4BN **Tel**: (UK and intl):+44 (0) 1444 475636 **E-mail**: ns.subs @qss-uk.com *or subscribe from their web site.*

Scientific American: *A monthly magazine containing specialist features. Articles range in level of reading difficulty and assumed knowledge.* Subscription enquiries: 415 Madison Ave. New York. NY10017-1111 **Tel**: (outside North America): 515-247-7631 **Tel**: (US& Canada): 800-333-1199

School Science Review: *A quarterly journal which includes articles, reviews, and news on current research and curriculum development. Free to Ordinary Members of the ASE or available on subscription.* Enquiries: **Tel**: 01707 28300 **Email**: info@ase.org.uk *or visit their web site.*

The American Biology Teacher: *The peer-reviewed journal of the NABT. Published nine times a year and containing information and activities relevant to biology teachers.* Contact: NABT, 12030 Sunrise Valley Drive, #110, Reston, VA 20191-3409 **Web**: www.nabt.org

The Origin and Evolution of Life

Describing the origin of life on Earth and the scientific evidence for evolution

The scientific evidence for evolution: the origins of life on Earth, the fossil record, comparative anatomy and biochemistry, biogeographical evidence.

Learning Objectives

☐ 1. Compile your own glossary from the **KEY WORDS** displayed in **bold type** in the learning objectives below.

The Origin of Life on Earth

The prebiotic world *(pages 10-13)*

☐ 2. Outline the conditions of **prebiotic Earth**, including reference to the role of the following: *high temperature, lightning, ultraviolet light penetration, and reducing atmosphere.* Explain the probable events that lead to the formation of life on Earth.

☐ 3. Provide a timeline for the major stages in the evolution of life on Earth. Summarise the main ideas related to where life originated: ocean surface, extraterrestrial (**panspermia**), and deep sea thermal vents.

☐ 4. Describe some of the geological and palaeontological evidence that suggests when life originated on Earth.

☐ 5. Outline the experiments (in particular the **Miller-Urey experiment**) that have attempted to simulate the **prebiotic environment** on Earth. Describe their importance in our understanding of the probable origin of organic compounds.

☐ 6. Discuss the hypothesis that the first catalysts responsible for **polymerisation reactions** were clay minerals and RNA.

The first cells *(pages 10-11, 14)*

☐ 7. Describe the possible role of RNA as the first self-replicating molecule. Discuss its role as an enzyme and its role in the origin of the first self-replicating cells.

☐ 8. Discuss the possible origin of membranes and the first prokaryotic cells.

☐ 9. Describe the evidence in the **geological record** for the first aquatic **prokaryotes**. Discuss the importance of these early organisms to the later evolution of diversity.

☐ 10. Distinguish between the **Eubacteria**, the **Archaea**, and the **Eukarya** with respect to their features and the environments in which they live. Explain what the current ecology of some bacterial groups tells us about the probable conditions of early life on Earth.

The origin of eukaryotes *(page 14)*

☐ 11. Recall the differences between eukaryotes and prokaryotes. Explain why the evolution of eukaryotic cells is regarded as a milestone in the development of complexity in living things.

☐ 12. Discuss the **endosymbiotic** (endosymbiont) **theory** for the evolution of eukaryotic cells. Summarise the evidence in support of this theory.

☐ 13. Summarise the main ideas about the evolution of multicellular life. Describe the benefits gained by the evolution of multicellularity (multicellular life).

The Evidence for Evolution

Background: The greatest obstacle to the establishment of evolutionary theory has been the difficulty in observing evolution in the time scales within which humans operate. Although more recently there have been direct observations made of populations evolving within observable time periods (flour beetles, bacteria, viruses, <u>Drosophila</u>), much of the evidence for evolution is indirect or circumstantial. The weight of accumulated evidence from many fields of science is overwhelmingly in support of evolution. The way in which organisms are classified reflects their evolutionary development (phylogeny) and degree of relatedness. Students should be aware that the scientific debate of evolution has centred on hypotheses for the evolutionary processes, <u>not</u> on the phenomenon of evolution itself.

The fossil record *(pages 15-21)*

☐ 14. State the conditions under which different **fossils** form. Include reference to **petrified remains**, **prints** and **moulds**, and preservation in **amber**, **tar**, **peat**, and **ice**.

☐ 15. Outline the methods for dating rocks and fossils using **radioisotopes**, e.g. ^{14}C and ^{40}K. Appreciate the degree of accuracy achieved by different dating methods and how the choice of isotope to use is made. Define the term: **half-life** and deduce the approximate age of materials based on a simple **decay curve** for a radioisotope.

☐ 16. Describe relative dating techniques using fossil sequence in strata. Distinguish between **relative dating** and **absolute dating** methods. Identify the different methods by which fossil remains are dated and describe when each of the dating methods is appropriate.

☐ 17. Appreciate that the dating of the main fossil-bearing rocks has provided the data for dividing the history of life on Earth into **geological periods**, which collectively form the geological time scale. Explain the system used to describe the age of rock strata (*era, period, epoch*).

☐ 18. Explain what is meant by **transitional fossils** and explain their significance. Offer an explanation for the apparent lack of transitional forms in the **fossil record**. Using examples, describe the trends that fossils indicate in the evolution of related groups.

☐ 19. Outline the **palaeontological evidence** for evolution using an example, e.g. evolution of horses or birds.

Comparative biochemistry *(pages 22-24)*

☐ 20. Explain the biochemical evidence by the universality of DNA, amino acids, and protein structures (e.g. cytochrome C) for the common ancestry of living organisms. Describe how comparisons of specific molecules between species are used as an indication of relatedness or phylogeny. Examples include comparisons of DNA, amino acid sequences, or blood proteins (see #21).

□ 21. Describe how **immunology** provides a method of quantifying the relatedness of species. Describe the basic principles and techniques involved.

□ 22. Discuss how biochemical variations can be used as an **evolutionary** (molecular) **clock** to determine probable dates of divergence from a common ancestor.

Anatomical comparisons *(pages 24-28)*

□ 23. In a general way, describe how **comparative anatomy**, **embryology**, and physiology have contributed to an understanding of evolutionary relationships.

□ 24. Distinguish between **homologous** structures and **analogous structures** arising as a result of convergent evolution. Give examples of **homology**. Explain the evidence for evolution provided by homologous anatomical structures, including: the vertebrate pentadactyl limb and vertebrate embryos. Note: Recognise that although vertebrate embryos may pass through similar stages during their development, ontogeny does not recapitulate phylogeny; Haeckel's original drawings were inaccurate and misleading.

□ 25. Explain how the contemporary field of evolutionary developmental biology (**evo-devo**) has provided some of the strongest evidence for the mechanisms of evolution, particularly for the evolution of novel forms. Appreciate that evo-devo compares the developmental processes of different organisms in an attempt to establish phylogenies and determine how developmental processes evolved.

□ 26. Discuss the significance of **vestigial organs** as indicators of evolutionary trends in some groups.

Biogeography *(pages 29-34, 47-48)*

□ 27. Using named examples, explain how the geographical distribution of plants and animals (both living and extinct), provides evidence of dispersal of organisms from a point of origin across pre-existing barriers.

□ 28. Outline the evidence for the occurrence of crustal movements by plate tectonics.

□ 29. EXTENSION: Describe an example of evolution occurring after physical isolation, e.g. Darwin's finches, invertebrates or parrots in New Zealand.

The Origin and Evolution of Life

Supplementary Texts

See page 7 for additional details of these texts:

■ Clegg, C.J., 1999. **Genetics and Evolution** (John Murray), pp. 60-65.

■ Futuyma, D.J., 2005. **Evolution**, (Sinauer Associates), chpt. 1-7 as required.

■ Jones, N., *et al.*, 2001. **The Essentials of Genetics**, pp. 202-205, 217-219.

■ Martin, R.A., 2004. **Missing Links**, chpt. 1 and case histories from section II as required.

■ Zimmer, C., 2001. **Evolution: The Triumph of an Idea**, (HarperCollins), chpt. 5-6 as required.

The following references for teachers provide detailed material on life's origins:

■ **The Molecular Origins of Life** (1998) Brack, A. (ed). Cambridge U.P. ISBN: 0-521-56475-1. *A thought provoking summary of this topic.*

■ **Biogenesis: Theories of Life's Origin** (1999) Lahav, N. Oxford University Press. ISBN: 0-19-511755-7. *A critical discussion of the study of the origin of life (detailed with good diagrams).*

Periodicals

See page 7 for details of publishers of periodicals:

STUDENT'S REFERENCE

■ **How Old is...** National Geographic, 200(3) September 2001, pp. 79-101. *A comprehensive discussion of dating methods and their application.*

■ **An RNA World** Biol. Sci. Rev., 11(3) January 1999, pp. 2-6. *An experiment to reproduce the prebiotic conditions on Earth suggests that RNA evolved before DNA as an early enzyme.*

■ **Other Worlds** New Scientist, 18 September 1999, pp. 24-47. *Articles in this special issue cover the conditions required for life to evolve and the organisation of the first biological life.*

■ **Primeval Pools** New Scientist, 2 July 2005, pp. 40-43. *An ecosystem where microbes still dominate as they did . of years ago*

■ **A Cool Early Life** Sci. American, Oct. 2005, pp. 40-47. *Discovery of ancient zircon crystals suggest that the earth cooled far sooner than once thought; as early as 4.4 billion ya. These cooler, wet surroundings were necessary for life to evolve.*

■ **The Rise of Life on Earth (series)** National Geographic, 193(3) March 1998, pp. 54-81. *The origins of life on Earth, the evolution of life's diversity, and the origin of eukaryotic cells.*

■ **A Waste of Space** New Scientist, 25 April 1998, pp. 38-39. *Vestigial organs: how they arise in an evolutionary sense and what role they may play.*

■ **The Quick and the Dead** New Scientist, 5 June 1999, pp. 44-48. *The formation of fossils: fossil types and preservation in different environments.*

■ **Life Grows Up** National Geographic, 193(4) April 1998, pp. 100-115. *The evolution of life: ancient fossils reveal the rise of life on Earth.*

■ **Meet your Ancestor** New Scientist, 9 Sept. 2006, pp. 35-39. *The significance of a recent fossil find: the missing link between fish and tetrapods.*

■ **Was Darwin Wrong?** National Geographic, 206(5) Nov. 2004, pp. 2-35. *Portrayal of the overwhelming scientific evidence for evolution.*

■ **Born Lucky** New Scientist, 12 July 2003, pp. 32-35. *This article discusses how, against odds, life established itself quickly on Earth, and suggests that this can tell us something about where life began. One of a series in this issue.*

■ **Proof of Life** New Scientist, 22 Feb. 2003, pp. 28-31. *New studies of microfossils suggest that life on Earth may be much younger than first thought.*

■ **The Golden Age of Dinosaurs** New Scientist, 21 May 2005, pp. 34-51. *A series of articles in a special issue exploring the discoveries that have transformed our understanding of dinosaurs..*

■ **Computers, DNA, and Evolution** Biol. Sci. Rev., 11(5) May 1999, pp. 24-29. *Using computers to compare the DNA sequences of different species and establish phylogeny.*

■ **Life's Rocky Start** Sci. American, April 2001, pp. 62-71. *The origins of life: prebiotic experiments & the role of minerals in early reactions on Earth.*

■ **The Ice of Life** Sci. American, August 2001, pp. 37-41. *Space ice may promote organic molecules and may have seeded life on Earth.*

■ **Earth in the Beginning** National Geographic, 210(6) Dec. 2006, pp. 58-67. *Modern landscapes offer glimpses of the way Earth may have looked billions of years ago.*

TEACHER'S REFERENCE

■ **Using Inquiry and Phylogeny to Teach Comparative Morphology** The American Biology Teacher, 67(6), Aug. 2005, pp. 412-417. *A hands-on, inquiry based approach to teaching comparative vertebrate skeletal morphology.*

■ **Astrobiology: Using Research to Investigate Science Curricula** The American Biology Teacher, 68(1), Jan. 2006, pp. 7-12. *The outline of a curriculum unit developed on astrobiology which focuses on current research into origins of life on earth and on other planets.*

■ **Putting Together Fossil Collections for 'Hands On' Evolution Laboratories** The American Biology Teacher, 63(1), January 2001, pp. 16-19. *How to put together a fossil lab: types of fossils (species) and contacts for material.*

■ **'New' Persuasive Evidence for Evolution** The American Biology Teacher, 60(9), Nov. 1998, pp. 662-70. *Investigating the molecular evidence for evolution: an account of how ancient 'errors' can persist in modern species.*

■ **Building a Phylogenetic Tree of the Human & Ape Superfamily Using DNA-DNA Hybridisation Data** The American Biology Teacher, 66(8), Oct. 2004, pp. 560-566. *A how-to-do-it activity determining genetic differences between species.*

■ **Haeckel's Embryos and Evolution** The American Biology Teacher, 61(5), May 1999, pp. 345-349. *An article that sets the record straight about the flaws in Haeckel's work. It examines how to study embryology in the context of phylogeny.*

Case study in transitional fossils: birds

■ **Winging It** New Scientist, 28 Aug. 1999, pp. 28-32. *Update on the evidence for the origin of bird flight.*

■ **Dinosaurs and Birds** The American Biology Teacher, 61(9), Nov. 1999, pp. 701-705. *A look at the bird-dinosaur link and the origins of flight.*

■ **Dinosaurs take Wing** National Geographic, 194(1) July 1998, pp. 74-99. *The evolution of birds from small theropod dinosaurs as evidenced by the homology between the typical dinosaur limb and the wing of the modern bird. An excellent article.*

Internet

See pages 4-5 for details of how to access **Bio Links** from our web site: **www.the biozone.com** From Bio Links, access sites under the topics:

EVOLUTION: • A history of evolutionary thought • Evolution • The Talk.Origins archive ... *and others* > **Evolution: Theory and Evidence**: • Evidence for evolution: an eclectic survey • Transitional vertebrate fossils FAQ ... *and others* > **The Fossil Record**: • Geological time scale • Geology and geologic time ... *and others* > **The Origins of Life on Earth:** • From primordial soup to prebiotic beach • Miller/Urey experiment • Origin of life • Origin of life on Earth

SPACE BIOLOGY > **Exobiology:** • Archaea in space • Cosmic ancestry • Evidence of primitive life from Mars • NASA's exobiology ... *and others*

Evolution **Presentation MEDIA** to support this topic: **EVOLUTION**

Life in the Universe

Life 'as we know it' requires three basic ingredients: a source of energy, carbon, and liquid water. Complex organic molecules (as are found in living things) have been detected beyond Earth in interstellar dust clouds and in meteorites that have landed on Earth. More than 4 billion years ago, one such dust cloud collapsed into a swirling **accretion disk** that gave rise to the sun and planets. Some of the fragile molecules survived the heat of solar system formation by sticking together in comets at the disk's fringe where temperatures were freezing. Later, the comets and other cloud remnants carried the molecules to Earth.

The formation of these organic molecules and their significance to the origin of life on Earth are currently being investigated experimentally (see below). The study of the origin of life on Earth is closely linked to the search for life elsewhere in our solar system. There are further plans to send solar and lunar orbiters to other planets and their moons, and even to land on a comet in November 2014. Their objective will be to look for signs of life (present or past) or its chemical precursors. If detected, such a discovery would suggest that life (at least 'primitive' life) may be widespread in the universe.

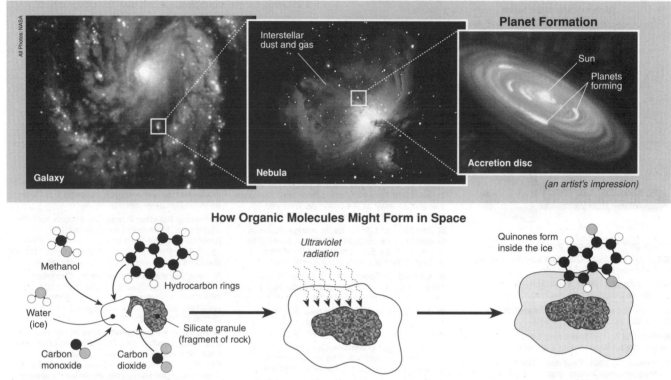

All Photos: NASA

Galaxy

Interstellar dust and gas

Nebula

Planet Formation

Sun

Planets forming

Accretion disc

(an artist's impression)

How Organic Molecules Might Form in Space

Methanol

Hydrocarbon rings

Water (ice)

Silicate granule (fragment of rock)

Carbon monoxide

Carbon dioxide

Ultraviolet radiation

Quinones form inside the ice

Interstellar ice begins to form when molecules such as methanol, water, and hydrocarbon freeze onto sandlike granules of silicate drifting in dense interstellar clouds.

Ultraviolet radiation from nearby stars cause some of the chemical bonds of the frozen compounds in the ice to break.

The broken down molecules recombine into structures such as quinones, which would never form if the fragments were free to float away.

NASA

Two **Mars Exploration rovers** landed on Mars in early 2004. Each rover carried sophisticated instruments, which were used to determine the history of climate and water at two sites where conditions may once have been favourable for life.

Organic Molecules Detected in Space

In a simple cloud-chamber experiment with simulated space ice (frozen water, methanol, and ammonia), complex compounds were yielded, including: ketones, nitriles, ethers, alcohols, and quinones (nearly identical in structure to those that help chlorophyll). These same organic molecules are found in carbon-rich meteorites. A six-carbon molecule (known as HMT) was also created. In warm, acidic water it is known to produce amino acids.

In another investigation into compounds produced in this way, some of the molecules displayed a tendency to form capsule-like droplets in water. These capsules were similar to those produced using extracts of a meteorite from Murchison, Australia in 1989. When organic compounds from the meteorite were mixed with water, they spontaneously assembled into spherical structures similar to cell membranes. These capsules were found to be made up of a host of complex organic molecules.

Source: *Life's far-flung raw materials*, Scientific American, July 1999, pp. 26-33

1. Suggest how sampling the chemical makeup of a comet might assist our understanding of life's origins:

_____

_____

2. Explain the significance of molecules from space that naturally form capsule-like droplets when added to water:

_____

_____

3. Explain how scientists are able to know about the existence of complex organic molecules in space: _____

_____

Code: A 3

The Origin of Life on Earth

Recent discoveries of **prebiotic** conditions on other planets and their moons has rekindled interest in the origin of life on primeval Earth. Experiments demonstrate that both peptides and nucleic acids may form polymers naturally in the conditions that are thought to have existed in a primitive terrestrial environment. RNA has also been shown to have enzymatic properties (**ribozymes**) and is capable of self-replication. These discoveries have removed some fundamental obstacles to creating a plausible scientific model for the origin of life from a prebiotic soup. Much research is now underway and space probes have been sent to Mercury, Venus, Mars, Pluto and its moon, Charon. They will search for evidence of prebiotic conditions or primitive microorganisms. The study of life in such regions beyond our planet is called **exobiology**.

Steps Proposed in the Origin of Life

The appearance of life on our planet may be understood as the result of evolutionary processes that involve the following major steps:

1. Formation of the Earth (4600 mya) and its acquisition of volatile organic chemicals by collision with comets and meteorites, which provided the precursors of biochemical molecules.

2. Prebiotic synthesis and accumulation of amino acids, purines, pyrimidines, sugars, lipids, and other organic molecules in the primitive terrestrial environment.

3. Prebiotic condensation reactions involving the synthesis of polymers of peptides (proteins), and nucleic acids (most probably just RNA) with self-replicating and catalytic (enzymatic) abilities.

4. Synthesis of lipids, their self-assembly into double-layered membranes and liposomes, and the 'capturing' of prebiotic (self-replicating and catalytic) molecules within their boundaries.

5. Formation of a **protobiont**; this is an immediate precursor to the first living systems. Such protobionts would exhibit cooperative interactions between small catalytic peptides, replicative molecules, proto-tRNA, and protoribosomes.

An RNA World

RNA has the ability to act as both genes and enzymes and offers a way around the "chicken-and-egg" problem: genes require enzymes to form; enzymes require genes to form. The first stage of evolution may have proceeded by RNA molecules performing the catalytic activities necessary to assemble themselves from a nucleotide soup. At the next stage, RNA molecules began to synthesise proteins. There is a problem with RNA as a prebiotic molecule because the ribose is unstable. This has led to the idea of a pre-RNA world (PNA).

Photo: Ron Lind

These living **stromatolites** from a beach in Western Australia are created by mats of bacteria. Similar, fossilised stromatolites have been found in rocks dating back to 3500 million years ago.

Dynamics of an RNA World

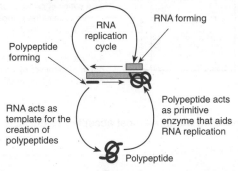

RNA replication cycle

RNA forming

Polypeptide forming

RNA acts as template for the creation of polypeptides

Polypeptide acts as primitive enzyme that aids RNA replication

Polypeptide

Scenarios for the Origin of Life

The origin of life remains a matter of scientific speculation. Three alternative views of how the key processes occurred are illustrated below:

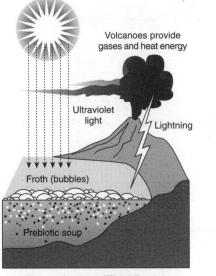

Volcanoes provide gases and heat energy

Ultraviolet light

Lightning

Froth (bubbles)

Prebiotic soup

Comet or meteorite from elsewhere in the solar system harbouring microorganisms

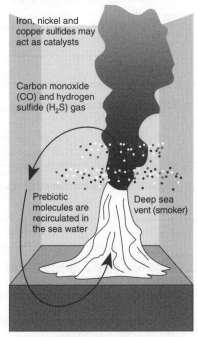

Iron, nickel and copper sulfides may act as catalysts

Carbon monoxide (CO) and hydrogen sulfide (H_2S) gas

Prebiotic molecules are recirculated in the sea water

Deep sea vent (smoker)

Ocean Surface (Tidal Pools)

This popular theory suggests that life arose in a tidepool, pond or on moist clay on the primeval Earth. Gases from volcanoes would have been energised by UV light or electrical discharges to form the prebiotic molecules in froth.

Panspermia

Cosmic ancestry (panspermia) is a serious scientific theory that proposes living organisms were 'seeded' on Earth as 'passengers' aboard comets and meteors. Such incoming organisms would have to survive the heat of re-entry.

Undersea Thermal Vents

A recently proposed theory suggests that life may have arisen at ancient volcanic vents (called smokers). This environment provides the necessary gases, energy, and a possible source of catalysts (metal sulfides).

Code: A 2

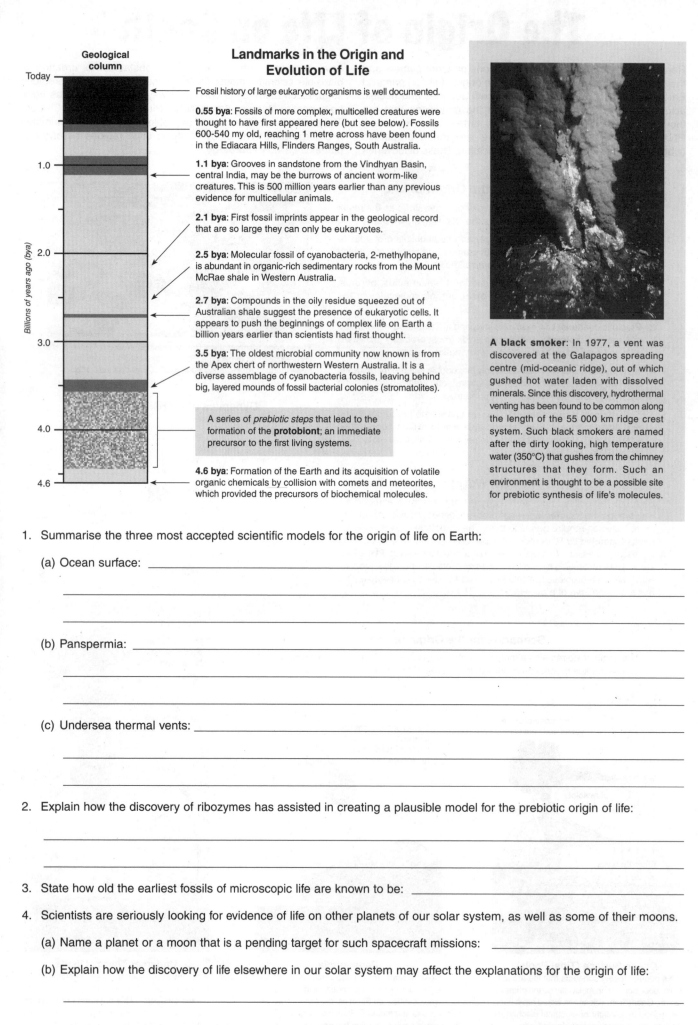

Geological column

Today

Billions of years ago (bya)

1.0

2.0

3.0

4.0

4.6

Landmarks in the Origin and Evolution of Life

Fossil history of large eukaryotic organisms is well documented.

0.55 bya: Fossils of more complex, multicelled creatures were thought to have first appeared here (but see below). Fossils 600-540 my old, reaching 1 metre across have been found in the Ediacara Hills, Flinders Ranges, South Australia.

1.1 bya: Grooves in sandstone from the Vindhyan Basin, central India, may be the burrows of ancient worm-like creatures. This is 500 million years earlier than any previous evidence for multicellular animals.

2.1 bya: First fossil imprints appear in the geological record that are so large they can only be eukaryotes.

2.5 bya: Molecular fossil of cyanobacteria, 2-methylhopane, is abundant in organic-rich sedimentary rocks from the Mount McRae shale in Western Australia.

2.7 bya: Compounds in the oily residue squeezed out of Australian shale suggest the presence of eukaryotic cells. It appears to push the beginnings of complex life on Earth a billion years earlier than scientists had first thought.

3.5 bya: The oldest microbial community now known is from the Apex chert of northwestern Western Australia. It is a diverse assemblage of cyanobacteria fossils, leaving behind big, layered mounds of fossil bacterial colonies (stromatolites).

A series of *prebiotic steps* that lead to the formation of the **protobiont**; an immediate precursor to the first living systems.

4.6 bya: Formation of the Earth and its acquisition of volatile organic chemicals by collision with comets and meteorites, which provided the precursors of biochemical molecules.

A black smoker: In 1977, a vent was discovered at the Galapagos spreading centre (mid-oceanic ridge), out of which gushed hot water laden with dissolved minerals. Since this discovery, hydrothermal venting has been found to be common along the length of the 55 000 km ridge crest system. Such black smokers are named after the dirty looking, high temperature water (350°C) that gushes from the chimney structures that they form. Such an environment is thought to be a possible site for prebiotic synthesis of life's molecules.

1. Summarise the three most accepted scientific models for the origin of life on Earth:

   (a) Ocean surface: _____

   _____

   _____

   (b) Panspermia: _____

   _____

   _____

   (c) Undersea thermal vents: _____

   _____

   _____

2. Explain how the discovery of ribozymes has assisted in creating a plausible model for the prebiotic origin of life:

   _____

   _____

3. State how old the earliest fossils of microscopic life are known to be: _____

4. Scientists are seriously looking for evidence of life on other planets of our solar system, as well as some of their moons.

   (a) Name a planet or a moon that is a pending target for such spacecraft missions: _____

   (b) Explain how the discovery of life elsewhere in our solar system may affect the explanations for the origin of life:

   _____

   _____

Prebiotic Experiments

In the 1950s, Stanley Miller and Harold Urey used equipment (illustrated below) to attempt to recreate the conditions on the primitive Earth. They hoped that the experiment might give rise to the biological molecules that were forerunners to the development of the first living organisms. Researchers at the time believed that the Earth's early atmosphere was made up of methane, water vapour, ammonia, and hydrogen gas. Many variations on this experiment, using a variety of recipes, have produced similar results. It seems that the building blocks of life are relatively easy to create. Many types of organic molecules have even been detected in deep space.

The Miller-Urey Experiment

The experiment (right) was run for a week after which samples were taken from the collection trap for analysis. Up to 4% of the carbon (from the methane) had been converted to amino acids. In this and subsequent experiments, it has been possible to form all 20 amino acids commonly found in organisms, along with nucleic acids, several sugars, lipids, adenine, and even ATP (if phosphate is added to the flask). Researchers now believe that the early atmosphere may be similar to the vapours given off by modern volcanoes: carbon monoxide (CO), carbon dioxide (CO_2), and nitrogen (N_2). Note the absence of free atmospheric oxygen.

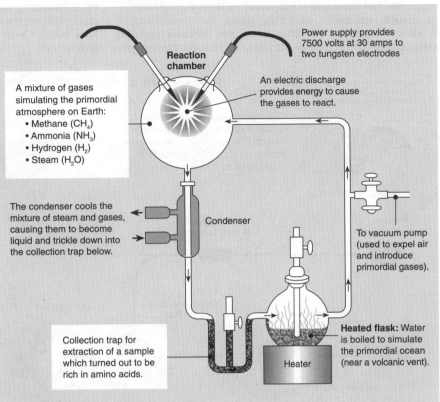

Power supply provides 7500 volts at 30 amps to two tungsten electrodes

Reaction chamber

An electric discharge provides energy to cause the gases to react.

A mixture of gases simulating the primordial atmosphere on Earth:
• Methane (CH_4)
• Ammonia (NH_3)
• Hydrogen (H_2)
• Steam (H_2O)

The condenser cools the mixture of steam and gases, causing them to become liquid and trickle down into the collection trap below.

Condenser

To vacuum pump (used to expel air and introduce primordial gases).

Collection trap for extraction of a sample which turned out to be rich in amino acids.

Heated flask: Water is boiled to simulate the primordial ocean (near a volcanic vent).

Heater

Iron pyrite, or 'fools gold' (above) has been proposed as a possible stabilising surface for the synthesis of organic compounds in the prebiotic world.

Some scientists envisage a global winter scenario for the formation of life. Organic compounds are more stable in colder temperatures and could combine in a lattice of ice. This frozen world could be thawed later.

Lightning is a natural phenomenon associated with volcanic activity. It may have supplied a source of electrical energy for the formation of new compounds (such as oxides of nitrogen) which were incorporated into organic molecules.

The early Earth was subjected to volcanism everywhere. At volcanic sites such as deep sea hydrothermal vents and geysers (like the one above), gases delivered vital compounds to the surface, where reactions took place.

1. In the Miller-Urey experiment simulating the conditions on primeval Earth, identify parts of the apparatus equivalent to:

    (a) Primeval atmosphere: _____

    (b) Primeval ocean: _____

    (c) Lightning: _____

    (d) Volcanic heat: _____

2. Name the organic molecules that were created by this experiment: _____

_____

3. (a) Suggest a reason why the Miller-Urey experiment is not an accurate model of what happened on the primeval Earth:

_____

    (b) Suggest changes to the experiment that could help it to better fit our understanding of the Earth's primordial conditions:

_____

The Origin and Evolution of Life

Code: A 3

The Origin of Eukaryotic Cells

The first firm evidence of eukaryotic cells is found in the fossil record at 540-600 mya. It is thought that eukaryotic cells evolved from large prokaryotic cells that ingested other free-floating prokaryotes. They formed a symbiotic relationship with the cells they engulfed (**endosymbiosis**). The two most important organelles that developed in eukaryotic cells were mitochondria, for aerobic respiration, and chloroplasts, for photosynthesis in aerobic conditions. Primitive eukaryotes probably acquired mitochondria by engulfing purple bacteria. Similarly, chloroplasts

may have been acquired by engulfing primitive cyanobacteria (which were already capable of photosynthesis). In both instances the organelles produced became dependent on the nucleus of the host cell to direct some of their metabolic processes. The sequence of evolutionary change shown below suggests that the lines leading to animal cells diverged before those leading to plant cells, but the reverse could also be true. Animal cells might have evolved from plant-like cells which subsequently lost their chloroplasts.

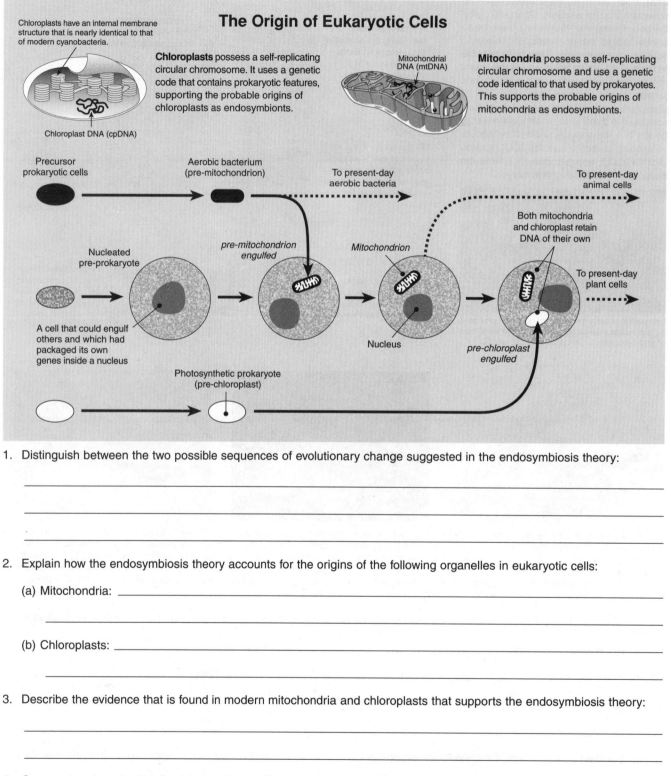

The Origin of Eukaryotic Cells

Chloroplasts have an internal membrane structure that is nearly identical to that of modern cyanobacteria.

Chloroplast DNA (cpDNA)

Chloroplasts possess a self-replicating circular chromosome. It uses a genetic code that contains prokaryotic features, supporting the probable origins of chloroplasts as endosymbionts.

Mitochondrial DNA (mtDNA)

Mitochondria possess a self-replicating circular chromosome and use a genetic code identical to that used by prokaryotes. This supports the probable origins of mitochondria as endosymbionts.

Precursor prokaryotic cells

Aerobic bacterium (pre-mitochondrion)

To present-day aerobic bacteria

To present-day animal cells

Nucleated pre-prokaryote

pre-mitochondrion engulfed

Mitochondrion

Both mitochondria and chloroplast retain DNA of their own

To present-day plant cells

A cell that could engulf others and which had packaged its own genes inside a nucleus

Nucleus

pre-chloroplast engulfed

Photosynthetic prokaryote (pre-chloroplast)

1. Distinguish between the two possible sequences of evolutionary change suggested in the endosymbiosis theory:

_____

_____

_____

2. Explain how the endosymbiosis theory accounts for the origins of the following organelles in eukaryotic cells:

   (a) Mitochondria: _____

   _____

   (b) Chloroplasts: _____

   _____

3. Describe the evidence that is found in modern mitochondria and chloroplasts that supports the endosymbiosis theory:

_____

_____

4. Comment on how the fossil evidence of early life supports or contradicts the endosymbiotic theory: _____

_____

_____

The History of Life on Earth

The scientific explanation of the origin of life on Earth is based soundly on the extensive fossil record, as well as the genetic comparison of modern life forms. Together they clearly indicate that modern life forms arose from ancient ancestors that have long since become extinct. These ancient life forms themselves originally arose from primitive cells living some 3500 million years ago in conditions quite different from those on Earth today. The earliest fossil records of living things show only simple cell types. It is believed that the first cells arose as a result of

evolution at the chemical level in a 'primordial soup' (a rich broth of chemicals in a warm pool of water, perhaps near a volcanic vent). Life appears very early in Earth's history, but did not evolve beyond the simple cell stage until much later, (about 600 mya). This would suggest that the development of complex life forms required more difficult evolutionary hurdles to be overcome. The buildup of free atmospheric oxygen, released as a by-product of photosynthesis, was important for the evolutionary development of animal life.

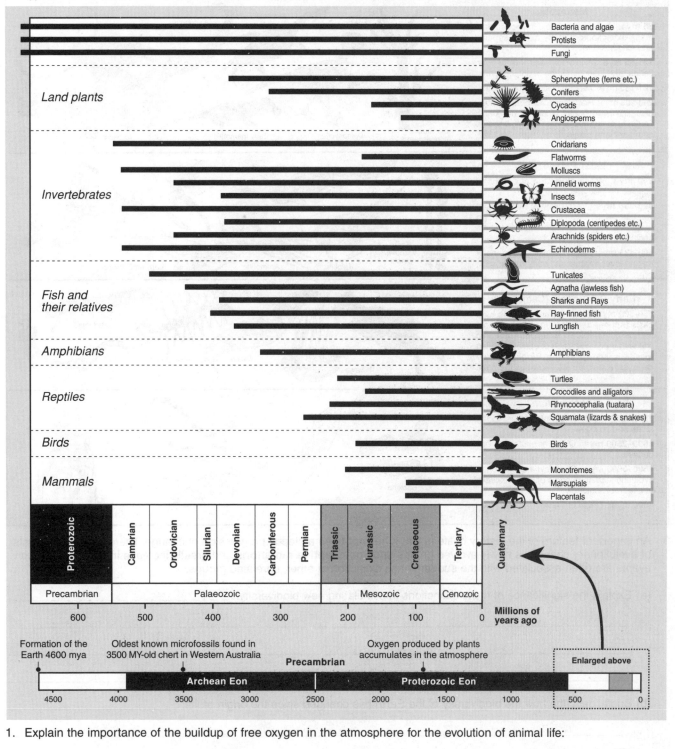

The Origin and Evolution of Life

1. Explain the importance of the buildup of free oxygen in the atmosphere for the evolution of animal life:

_____

2. Using the diagram above, determine how many millions of years ago the fossil record shows the first appearance of:

(a) Invertebrates: _____   (b) Fish (ray-finned): _____   (c) Land plants: _____

(d) Reptiles: _____   (e) Birds: _____   (f) Mammals: _____

Cenozoic

1.65 mya: Modern humans evolve and their hunting activities, starting at the most recent ice age, cause the most recent mass extinction.

3-5 mya: Early humans arise from ape ancestors.

65-1.65 mya: Major shifts in climate. Major adaptive radiations of angiosperms (flowering plants), insects, birds and mammals.

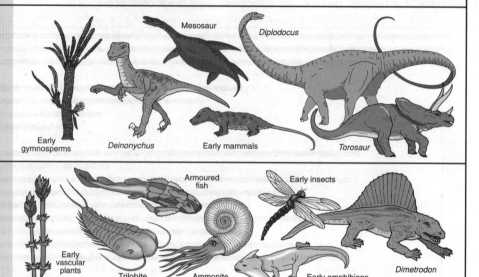

Unitatherium
Diatryma
Deinotherium
Glyptodon
Humans
Sabre-tooth cats

Mesozoic

65 mya: Apparent asteroid impact causes mass extinctions of many marine species and all dinosaurs.

135-65 mya: Major radiations of dinosaurs, fishes, and insects. Origin of angiosperms.

181-135 mya: Major radiations of dinosaurs.

240-205 mya: Recoveries, adaptive radiation of marine invertebrates, dinosaurs and fishes. Origin of mammals Gymnosperms become dominant land plants.

Mesosaur
Diplodocus
Early gymnosperms
Deinonychus
Early mammals
Torosaur

Later Palaeozoic

240 mya: Mass extinction of nearly all species on land and in the sea.

435-280 mya: Vast swamps with the first vascular plants. Origin and adaptive radiation of reptiles, insects and spore bearing plants (including gymnosperms).

500-435 mya: Major adaptive radiations of marine invertebrates and early fishes.

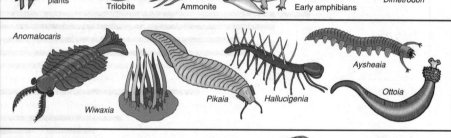

Armoured fish
Early insects
Early vascular plants
Trilobite
Ammonite
Early amphibians
Dimetrodon

Early Palaeozoic (Cambrian)

550-500 mya: Origin of animals with hard parts (appear as fossils in rocks). Simple marine communities. A famous Canadian site with a rich collection of early Cambrian fossils is known as the Burgess Shale deposits; examples are shown on the right.

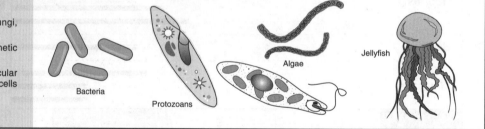

Anomalocaris
Aysheaia
Ottoia
Wiwaxia
Pikaia
Hallucigenia

Precambrian

2500–570 mya: Origin of protists, fungi, algae and animals.

3800–2500 mya: Origin of photosynthetic bacteria.

4600–3800 mya: Chemical and molecular evolution leading to origin of life; protocells to anaerobic bacteria.

4600 mya: Origin of Earth.

Bacteria
Protozoans
Algae
Jellyfish

3. An important feature of the history of life is that it has not been a steady progression of change. There have been bursts of evolutionary change as newly evolved groups undergo **adaptive radiations** and greatly increase in biodiversity. Such events are often associated with the sudden mass extinction of other, unrelated groups.

(a) Explain the significance of mass extinctions in stimulating new biodiversity: _____

_____

_____

_____

(b) Briefly describe how the biodiversity of the Earth has changed since the origin of life: _____

_____

_____

_____

_____

_____

Fossil Formation

Fossils are the remains of long-dead organisms that have escaped decay and have, after many years, become part of the Earth's crust. A fossil may be the preserved remains of the organism itself, the impression of it in the sediment (cast), or marks made by it during its lifetime (called trace fossils). For fossilisation to occur, rapid burial of the organism is required (usually in water-borne sediment). This is followed by chemical alteration, where minerals are added or removed. Fossilisation requires the normal processes of decay to be permanently arrested. This can occur if the remains are isolated from the air or water and decomposing microbes are prevented from breaking them down. Fossils provide a record of the appearance and extinction of organisms, from species to whole taxonomic groups. Once this record is calibrated against a time scale (by using a broad range of dating techniques), it is possible to build up a picture of the evolutionary changes that have taken place.

Modes of Preservation

Silicification: Silica from weathered volcanic ash is gradually incorporated into partly decayed wood (also called petrification).

Phosphatisation: Bones and teeth are preserved in phosphate deposits.

Pyritisation: Iron pyrite replaces hard remains of the dead organism.

Tar pit: Animals fall into and are trapped in mixture of tar and sand.

Trapped in amber: Gum from conifers traps insects and then hardens.

Limestone: Calcium carbonate from the remains of marine plankton is deposited as a sediment that traps the remains of other sea creatures.

Brachiopod (lamp shell), Jurassic (New Zealand)

Cast: This impression of a lamp shell is all that is left after the original shell material was dissolved after fossilisation.

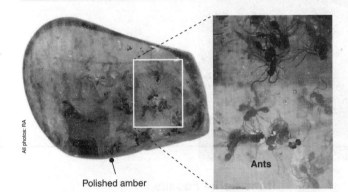

All photos: RA

Polished amber

Ants

Insects in amber: The fossilised resin or gum produced by some ancient conifers trapped these insects (including the ants visible in the enlargement) about 25 million years ago (Madagascar).

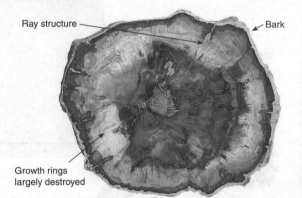

Ray structure

Bark

Growth rings largely destroyed

Petrified wood: A cross-section of a limb from a coniferous tree (Madagascar).

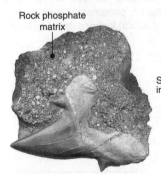

Rock phosphate matrix

Shark tooth: The tooth of a shark *Lamna obliqua* from phosphate beds, Eocene (Khouribga, Morocco).

Shell

Stone interior

Ammonite: This ammonite still has a layer of the original shell covering the stone interior, Jurassic (Madagascar).

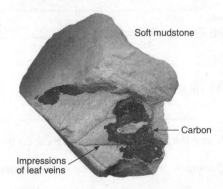

Sand and tar matrix

Wing bones

Bird bones: Fossilised bones of a bird that lived about 5 million years ago and became stuck in the tar pits at la Brea, Los Angeles, USA.

Shell and chambers replaced by iron pyrite

Ammonite: This ammonite has been preserved by a process called pyritisation, late Cretaceous (Charmouth, England).

Fossil fern: This compression fossil of a fern frond shows traces of carbon and wax from the original plant, Carboniferous (USA).

Soft mudstone

Carbon

Impressions of leaf veins

Sub-fossil: Leaf impression in soft mudstone (can be broken easily with fingers) with some of the remains of the leaf still intact (a few thousand years old, New Zealand).

The Origin and Evolution of Life

Code: A 1

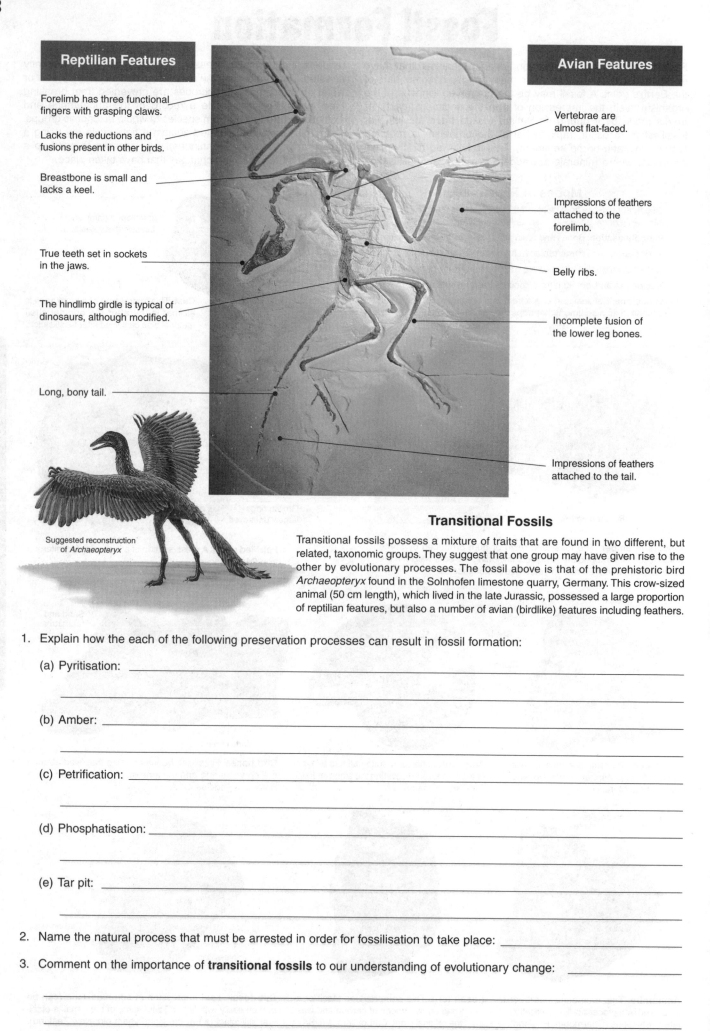

Reptilian Features

Forelimb has three functional fingers with grasping claws.

Lacks the reductions and fusions present in other birds.

Breastbone is small and lacks a keel.

True teeth set in sockets in the jaws.

The hindlimb girdle is typical of dinosaurs, although modified.

Long, bony tail.

Suggested reconstruction of *Archaeopteryx*

Avian Features

Vertebrae are almost flat-faced.

Impressions of feathers attached to the forelimb.

Belly ribs.

Incomplete fusion of the lower leg bones.

Impressions of feathers attached to the tail.

Transitional Fossils

Transitional fossils possess a mixture of traits that are found in two different, but related, taxonomic groups. They suggest that one group may have given rise to the other by evolutionary processes. The fossil above is that of the prehistoric bird *Archaeopteryx* found in the Solnhofen limestone quarry, Germany. This crow-sized animal (50 cm length), which lived in the late Jurassic, possessed a large proportion of reptilian features, but also a number of avian (birdlike) features including feathers.

1. Explain how the each of the following preservation processes can result in fossil formation:

   (a) Pyritisation: _____

   _____

   (b) Amber: _____

   _____

   (c) Petrification: _____

   _____

   (d) Phosphatisation: _____

   _____

   (e) Tar pit: _____

   _____

2. Name the natural process that must be arrested in order for fossilisation to take place: _____

3. Comment on the importance of **transitional fossils** to our understanding of evolutionary change: _____

   _____

   _____

The Fossil Record

The diagram below represents a cutting into the earth revealing the layers of rock. Some of these layers may have been laid down by water (sedimentary rocks) or by volcanic activity (volcanic rocks). Fossils are the actual remains or impressions of plants or animals that become trapped in the sediments after their death. Layers of sedimentary rock are arranged in the order that they were deposited, with the most recent layers near the surface (unless they have been disturbed).

Profile with Sedimentary Rocks Containing Fossils

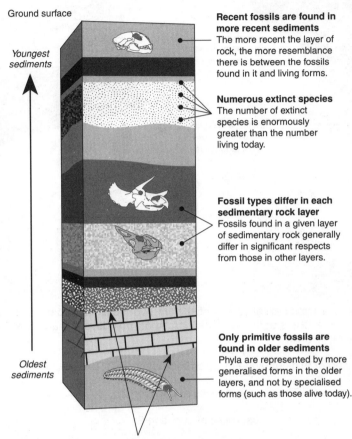

Ground surface

Youngest sediments

Oldest sediments

Recent fossils are found in more recent sediments
The more recent the layer of rock, the more resemblance there is between the fossils found in it and living forms.

Numerous extinct species
The number of extinct species is enormously greater than the number living today.

Fossil types differ in each sedimentary rock layer
Fossils found in a given layer of sedimentary rock generally differ in significant respects from those in other layers.

Only primitive fossils are found in older sediments
Phyla are represented by more generalised forms in the older layers, and not by specialised forms (such as those alive today).

New fossil types mark changes in environment
In the rocks marking the end of one geological period, it is common to find many new fossils that become dominant in the next. Each geological period had an environment very different from those before and after. Their boundaries coincided with drastic environmental changes and the appearance of new niches. These produced new selection pressures resulting in new adaptive features in the surviving species, as they responded to the changes.

The rate of evolution can vary

According to the fossil record, rates of evolutionary change seem to vary. There are bursts of species formation and long periods of relative stability within species (stasis). The occasional rapid evolution of new forms apparent in the fossil record, is probably a response to a changing environment. During periods of stable environmental conditions, evolutionary change may slow down.

The Fossil Record of Proboscidea

African and Indian elephants have descended from a diverse group of animals known as **proboscideans** (named for their long trunks). The first pig-sized, trunkless members of this group lived in Africa 40 million years ago. From Africa, their descendants invaded all continents except Antarctica and Australia. As the group evolved, they became larger; an effective evolutionary response to deter predators. Examples of extinct members of this group are illustrated below:

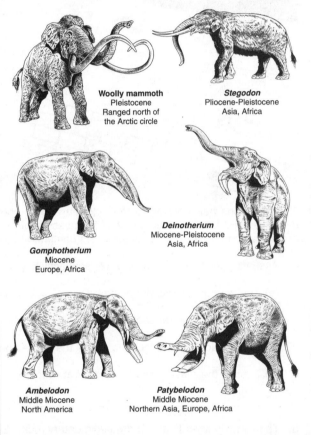

Woolly mammoth
Pleistocene
Ranged north of the Arctic circle

Stegodon
Pliocene-Pleistocene
Asia, Africa

Gomphotherium
Miocene
Europe, Africa

Deinotherium
Miocene-Pleistocene
Asia, Africa

Ambelodon
Middle Miocene
North America

Patybelodon
Middle Miocene
Northern Asia, Europe, Africa

- **Modern day species can be traced:** The evolution of many present-day species can be very well reconstructed. For instance, the evolutionary history of the modern elephants is exceedingly well documented for the last 40 million years. The modern horse also has a well understood fossil record spanning the last 50 million years.

- **Fossil species are similar to but differ from today's species:** Most fossil animals and plants belong to the same major taxonomic groups as organisms living today. However, they do differ from the living species in many features.

1. Name an animal or plant taxon (e.g. family, genus, or species) that has:

   (a) A good fossil record of evolutionary development: _____

   (b) Appeared to have changed very little over the last 100 million years or so: _____

2. Discuss the importance of **fossils** as a record of evolutionary change over time: _____

   _____

   _____

   _____

The Origin and Evolution of Life

Code: RA 2

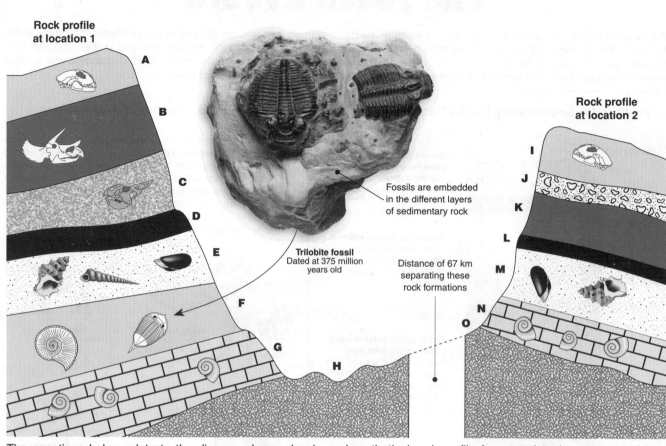

Rock profile at location 1

A
B
C
D
E
F
G
H

Fossils are embedded in the different layers of sedimentary rock

Trilobite fossil
Dated at 375 million years old

Distance of 67 km separating these rock formations

Rock profile at location 2

I
J
K
L
M
N
O

The questions below relate to the diagram above, showing a hypothetical rock profile from two locations separated by a distance of 67 km. There are some differences between the rock layers at the two locations. Apart from layers D and L which are volcanic ash deposits, all other layers comprise sedimentary rock.

3. Assuming there has been no geological activity (e.g. tilting or folding), state in which rock layer (A-O) you would find:

   (a) The youngest rocks at Location 1: _____   (c) The youngest rocks at Location 2: _____

   (b) The oldest rocks at Location 1: _____   (d) The oldest rocks at Location 2: _____

4. (a) State which layer at location 1 is of the same age as layer M at location 2: _____

   (b) Explain the reason for your answer above: _____

   _____

5. The rocks in layer H and O are sedimentary rocks. Explain why there are no visible fossils in layers:

   _____

   _____

6. (a) State which layers present at location 1 are missing at location 2: _____

   (b) State which layers present at location 2 are missing at location 1: _____

7. Describe three methods of dating rocks: _____

   _____

   _____

8. Using radiometric dating, the trilobite fossil was determined to be approximately 375 million years old. The volcanic rock layer (D) was dated at 270 million years old, while rock layer B was dated at 80 million years old. Give the approximate **age range** (i.e. greater than, less than or between given dates) of the rock layers listed below:

   (a) Layer A: _____   (d) Layer G: _____

   (b) Layer C: _____   (e) Layer L: _____

   (c) Layer E: _____   (f) Layer O: _____

Dating Fossils

Fossils are rarely able to be dated directly. In general, it is the rocks in which they are found that are dated. The exception is radiocarbon dating, which can directly measure the age of the organic matter in a sample. Dating usually begins with an attempt to order past events in a rock profile, and to relate the fossils to datable rock layers. Many techniques for measuring the age of rocks and minerals have been established. In the early days of developing these techniques there were problems in producing dependable results, but the methods have been much refined and often now provide dates with a high degree of certainty. Multiple dating methods may be applied to samples, providing cross-referencing, which gives further confidence in a given date. Dating methods can be grouped into two categories: those that rely on the gradual radioactive decay of an element (e.g. **radiocarbon**, **potassium-argon**, **fission track**); and those that use other methods (e.g. **tree-rings**, **palaeomagnetism**).

Dating Method	Useable Dating Range (years)	Datable Materials
Methods using radioisotopes:	100 million 10 million 1 million 100 000 10 000 1000 (Log scale)	
Fission track: Uranium (^{235}U) sometimes undergoes spontaneous fission, and the subatomic particles emitted leave tracks through the mineral.		Pottery, glass, and volcanic minerals.
Radiocarbon (^{14}C): Measures the loss of the isotope *carbon-14,* taken up by an organism when it was alive, within its fossilised remains.		Wood, shells, peat, charcoal, bone, animal tissue, calcite, soil.
Potassium/Argon (K/Ar): Measures the decay of *potassium-40* to *argon-40* in volcanic rocks that lie above or below fossil bearing strata.		Volcanic rocks and minerals.
Uranium series: Measures the decay of the two main isotopes of uranium (^{235}U and ^{238}U) into thorium (^{230}Th) and another isotope (^{234}U) respectively.		Marine carbonate, coral, mollusc shells
Non-isotopic methods:		
Palaeomagnetism: Shows the alignment of the Earth's magnetic field at the time when the rock sample was last heated above a critical level.		Rocks that contain iron-bearing minerals
Thermoluminescence: Measures the light emitted by a sample of quartz and/or zircon grains that has been exposed to sunlight or fire in the distant past.		Ceramics, quartz, feldspar, carbonates
Electron spin resonance (ESR): Measures the microwave energy absorbed by samples previously heated or exposed to sunlight in the distant past.		Burnt flints, cave sediments, bone, teeth, loess (wind-blown deposits)
Amino acid racemisation: Measures the gradual conversion of left- to right-handed amino acid isomers in the proteins preserved in organic remains.		Organic remains
Varve: Measures the distinct, annually deposited layers of sediments (varves) found in many lakes.		Mainly glacial lakes
Tree-ring: Measures the annual growth rings of trees (can be cross referenced with C-14 dating).		Trees, timber from buildings, ships

The Origin and Evolution of Life

1. Examine the diagram above and determine the approximate dating range (note the logarithmic time scale) and datable materials for each of the methods listed below:

                            Dating Range   **Datable Materials**

   (a) Potassium-argon method: _____   _____

   (b) Radiocarbon method: _____   _____

   (c) Tree-ring method: _____   _____

   (d) Thermoluminescence: _____   _____

2. When the date of a sample has been determined, it is common practice to express it in the following manner:
Example: **1.88 ± 0.02** million years old. Explain what the **± 0.02** means in this case:

_____

3. Suggest a possible source of error that could account for an incorrect dating measurement using a radioisotope method:

_____

DNA Hybridisation

The more closely two species are related, the fewer differences there will be in the exact sequence of bases. This is because there has been less time for the point mutations that will bring about these changes to occur. Modern species can be compared to see how long ago they shared a **common ancestor.** This technique gives a measure of 'relatedness', and can be calibrated against known fossil dates to create a **molecular clock**. It is then possible to give approximate dates of common origin to species with no or poor fossil data. This method has been applied to primate DNA samples to help determine the approximate date of human divergence from the apes, which has been estimated to be between 10 and 5 million years ago.

DNA Hybridisation

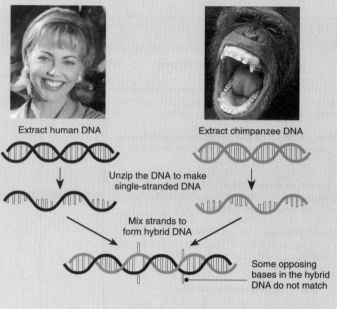

Extract human DNA   Extract chimpanzee DNA

Unzip the DNA to make single-stranded DNA

Mix strands to form hybrid DNA

Some opposing bases in the hybrid DNA do not match

1. Blood samples from each species are taken, from which the DNA is isolated.

2. The DNA from each species is made to unwind into single strands by applying heat (both human and chimpanzee DNA unwinds at 86°C).

3. Enzymes are used to snip the single strands of DNA into smaller pieces (about 500 base pairs long).

4. The segments from human and chimpanzee DNA are combined to see how closely they bind to each other (single strand segments tend to find their complementary segments and rewind into a double helix again).

5. The greater the similarity in DNA base sequence, the stronger the attraction between the two strands and therefore they are harder to separate again. By measuring how hard this hybrid DNA is to separate, a crude measure of DNA 'relatedness' can be achieved.

6. The degree of similarity of the hybrid DNA can be measured by finding the temperature that it unzips into single strands again (in this case it would be 83.6°C).

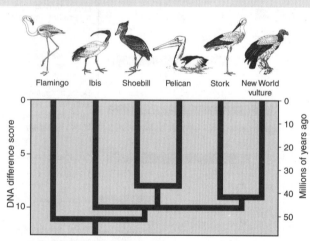

The relationships among the **New World vultures** and **storks** have been determined using DNA hybridisation. It has been possible to estimate how long ago various members of the group shared a common ancestor.

Similarity of human DNA to that of other primates

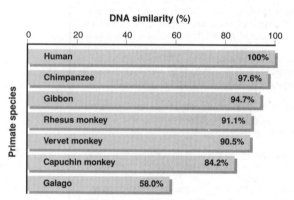

DNA similarity (%)

Primate species	DNA similarity
Human	100%
Chimpanzee	97.6%
Gibbon	94.7%
Rhesus monkey	91.1%
Vervet monkey	90.5%
Capuchin monkey	84.2%
Galago	58.0%

The genetic relationships among the **primates** has been investigated using DNA hybridisation. Human DNA was compared with that of the other primates. It largely confirmed what was suspected from anatomical evidence.

1. Explain how **DNA hybridisation** can give a measure of genetic relatedness between species:

_____

_____

_____

2. Study the graph showing the results of a DNA hybridisation between human DNA and that of other primates.

   (a) State which is the most closely related primate to humans: _____

   (b) State which is the most distantly related primate to humans: _____

3. State the DNA difference score for: (a) Shoebills and pelicans: _____   (b) Storks and flamingos: _____

4. On the basis of DNA hybridisation, state how long ago the ibises and New World vultures shared a common ancestor:

_____

Code: DA 2

Immunological Studies

Immunological studies provide a method of indirectly estimating the degree of similarity of proteins in different species. If differences exist in the proteins, then there must also be differences in the DNA that codes for them. The evolutionary relationships of a large number of different animal groups have been established on the basis of immunology. The results support the phylogenies developed from other areas: biogeography, comparative anatomy, and fossil evidence.

Method for Immunological Comparison

1. Blood serum (containing blood proteins but no cells) is collected from a human and is injected into a rabbit. This causes the formation of antibodies in the rabbit's blood. These identify human blood proteins, attach to them and render them harmless.

2. A sample of the rabbit's blood is taken and the rabbit's antibodies that recognise human blood proteins are extracted.

3. These anti-human antibodies are then added to blood samples from other species to see how well they recognise the proteins in the different blood. The more similar the blood sample is to original human blood, the greater the reaction (which takes the form of creating a precipitate, i.e. solids).

The five blood samples that were tested (on the right) show varying degrees of precipitate (solid) formation. Note that when the anti-human antibodies are added to human blood there is a high degree of affinity. There is poor recognition when added to rat blood.

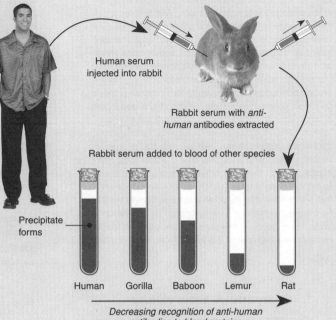

Human serum injected into rabbit

Rabbit serum with *anti-human* antibodies extracted

Rabbit serum added to blood of other species

Precipitate forms

Human   Gorilla   Baboon   Lemur   Rat

Decreasing recognition of anti-human antibodies to blood proteins →

Immunological Comparison of Tree Frogs

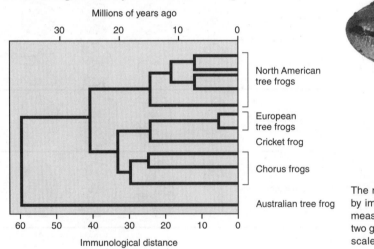

Millions of years ago

30   20   10   0

North American tree frogs

European tree frogs

Cricket frog

Chorus frogs

Australian tree frog

60   50   40   30   20   10   0

Immunological distance

The relationships among **tree frogs** have been established by immunological studies. The immunological distance is a measure of the number of amino acid substitutions between two groups. This, in turn, has been calibrated to provide a time scale showing when the various related groups diverged.

1. Briefly describe how **immunological studies** have contributed evidence that the process of evolution has taken place:

_____

_____

_____

_____

2. Study the graph above showing the immunological distance between tree frogs. State the immunological distance between the following frogs:

   (a) Cricket frog and the Australian tree frog: _____ (b) The various chorus frogs: _____

3. Describe how closely the Australian tree frog is related to the other frogs shown:

_____

4. State when the North American tree frogs became separated from the European tree frogs: _____

Other Evidence for Evolution

Amino Acid Sequences

Each of our proteins has a specific number of amino acids arranged in a specific order. Any differences in the sequence reflect changes in the DNA sequence. The haemoglobin beta chain has been used as a standard molecule for comparing the precise sequence of amino acids in different species. Haemoglobin is the protein in our red blood cells that is responsible for carrying oxygen around our bodies. The haemoglobin in adults is made up of four polypeptide chains: 2 alpha chains and 2 beta chains. Each is coded for by a separate gene.

Example right: When the sequence of human haemoglobin, which is 146 amino acids long, was compared with that of 5 other primate species it was found that chimpanzees had an identical sequence while those that were already considered less closely related had a greater number of differences. This suggests a very close genetic relationship between humans, chimpanzees and gorillas, but less with the other primates.

Amino Acid Differences Between Humans and Other Primates

The *'position of changed amino acid'* is the point in the protein, composed of 146 amino acids, at which the **different** amino acids occurs

Primate	No. of amino acids different from humans	Position of changed amino acids
Chimpanzee	Identical	–
Gorilla	1	104
Gibbon	3	80 87 125
Rhesus monkey	8	9 13 33 50 76 87 104 125
Squirrel monkey	9	5 6 9 21 22 56 76 87 125

Comparative Embryology

By comparing the development of embryos from different species, Ernst von Bayer in 1828 noticed that animals are more similar during early stages of their embryological development than later as adults. This later led to Ernst Haeckel (1834-1919) to propose his famous principle: *ontogeny recapitulates phylogeny*. He claimed that the development of an individual (ontogeny) retraces the stages through which the individual species has passed during its evolution (phylogeny). This idea is now known to be an oversimplification and is misleading. Although early developmental sequences between all vertebrates are similar, there are important deviations from the general developmental plan in different species. Notice the gill slits that briefly appear in the human embryo (arrowed). The more closely related forms of the monkey and humans continue to appear similar until a later stage in development, compared to more distantly related species. From the study of foetal development it is possible to find clues as to how evolution generates the diversity of life forms through time, but 'ontogeny does not recapitulate phylogeny'.

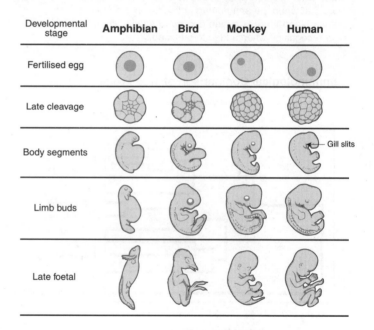

1. Study the table of data showing the differences in **amino acid sequences** for selected primates. Explain why chimpanzees and gorillas are considered most closely related to humans, while monkeys are less so:

_____

_____

_____

2. Briefly describe how **comparative embryology** has contributed evidence to support the concept of evolution:

_____

_____

3. Describe a commonly used biochemical method for precisely analysing the genes in organisms to determine their evolutionary relationships:

_____

_____

The Evolution of Novel Forms

The relatively new field of **evolutionary developmental biology** (or evo-devo) addresses the origin and evolution of embryonic development and looks at how modifications of developmental processes can lead to novel features. Scientists now know that specific genes in animals, including a subgroup of genes called *Hox* genes, are part of a basic **'tool kit'** of genes that control animal development. Genomic studies have shown that these genes are **highly conserved** (i.e. they show little change in different lineages). Very disparate organisms share the same **tool kit** of genes, but regulate them differently. The implication of this is that large changes in morphology or function are associated with changes in gene regulation, rather than the evolution of new genes, and natural selection associated with gene switches plays a major role in evolution.

The Role of *Hox* Genes

Hox genes control the development of back and front parts of the body. The same genes (or homologous ones) are present in essentially all animals, including humans.

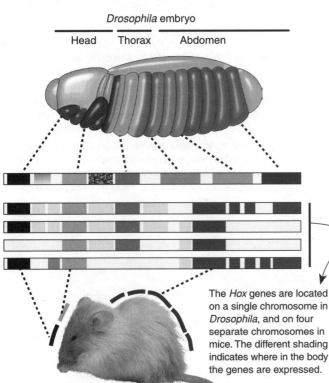

Drosophila embryo

Head   Thorax   Abdomen

The *Hox* genes are located on a single chromosome in *Drosophila*, and on four separate chromosomes in mice. The different shading indicates where in the body the genes are expressed.

The Evolution of Novel Forms

Even very small changes (mutations) in the *Hox* genes can have a profound effect on morphology. Such changes to the genes controlling development have almost certainly been important in the evolution of novel structures and body plans. Four principles underly the evo-devo thinking regarding the evolution of novel forms:

- **Evolution works with what is already present**: New structures are modifications of pre-existing structures.

- **Multifunctionality** and **redundancy**: Functional redundancy in any part of a multifunctional structure allows for specialisation and division of labour through the development of two separate structures.

  Example: the diversity of appendages (including mouthparts) in arthropods.

- **Modularity**: Modular architecture in animals (arthropods, vertebrates) allows for the modification and specialisation of individual body parts. Genetic switches allow changes in one part of a structure, independent of other parts.

The Origin and Evolution of Life

Shifting *Hox* Expression

Huge diversity in morphology in organisms within and across phyla could have arisen through small changes in the genes controlling development.

Differences in neck length in vertebrates provides a good example of how changes in gene expression can bring about changes in morphology. Different vertebrates have different numbers of neck vertebrae. The boundary between neck and trunk vertebrae is marked by expression of the **Hox c6 gene** (c6 denotes the sixth cervical or neck vertebra) in all cases, but the position varies in each animal relative to the overall body. The forelimb (arrow) arises at this boundary in all four-legged vertebrates. In snakes, the boundary is shifted forward to the base of the skull and no limbs develop. As a result of these differences in expression, mice have a short neck, geese a long neck, and snakes, no neck at all.

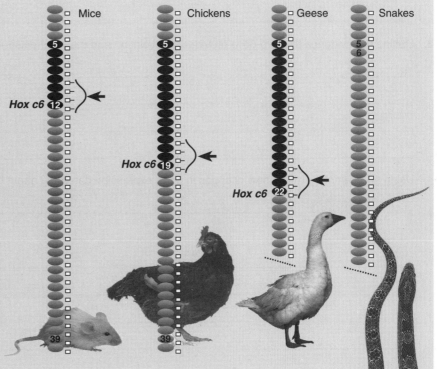

Mice   Chickens   Geese   Snakes

Hox c6   12

Hox c6   19

Hox c6   22

Code: A 3

Genetic Switches in Evolution

The *Hox* genes are just part of the collection of genes that make up the genetic tool kit for animal development. The genes in the tool kit act as switches, shaping development by affecting how other genes are turned on or off. The distribution of genes in the tool kit indicates that it is ancient and was in place before the evolution of most types of animals. Differences in form arise through changes in genetic switches. One example is the evolution of eyespots in butterflies:

■ The ***Distal-less*** gene is one of the important **master body-building genes** in the genetic tool kit. Switches in the *Distal-less* gene control expression in the embryo (E), larval legs (L), and wing (W) in flies and butterflies, but butterflies have also evolved an extra switch (S) to control eyespot development.

■ Once *Distal-less* spots evolved, changes in *Distal-less* expression (through changes in the switch) produced more or fewer spots.

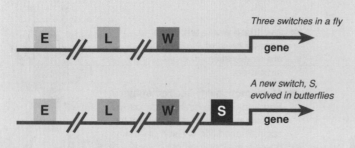

Changes in Distal-less regulation were probably achieved by changing specific sequences of the Distal-less gene eyespot switch. The result? Changes in eyespot size and number.

Same Gene, New Tricks

Stichophthalma camadeva

Junonia coenia (buckeye)

Taenaris macrops

■ The action of a tool kit protein depends on context: where particular cells are located at the time when the gene is switched on.

■ Changes in the DNA sequence of a genetic switch can change the zone of gene expression without disrupting the function of the tool kit protein itself.

■ The spectacular **eyespots** on butterfly wings (arrowed above) represent different degrees of a basic pattern, from virtually all eyespot elements expressed (*Stichophthalma*) to very few (*Taenaris*).

1. Explain what is meant by "evo-devo" and explain its aims: _____

_____

_____

2. Briefly describe the role of *Hox* genes in animal development: _____

_____

_____

_____

3. Outline the evidence that evo-devo provides for evolution and the mechanisms by which it occurs: _____

_____

_____

_____

_____

4. Using an example, discuss how changes in gene expression can bring about changes in morphology:

_____

_____

_____

_____

_____

_____

_____

Comparative Anatomy

The evolutionary relationships between groups of organisms is determined mainly by structural similarities called **homologous structures** (homologies), which suggest that they all descended from a common ancestor with that feature. The bones of the forelimb of air-breathing vertebrates are composed of similar bones arranged in a comparable pattern. This is indicative of a common ancestry. The early land vertebrates were amphibians and possessed a limb structure called the **pentadactyl limb**: a limb with 5 fingers or toes (below left). All vertebrates that descended from these early amphibians, including reptiles, birds and mammals, have limbs that have evolved from this same basic pentadactyl pattern. They also illustrate the phenomenon known as **adaptive radiation**, since the basic limb plan has been adapted to meet the requirements of different niches.

Generalised Pentadactyl Limb

The forelimbs and hind limbs have the same arrangement of bones but they have different names. In many cases bones in different parts of the limb have been highly modified to give it a specialised locomotory function.

Specialisations of Pentadactyl Limbs

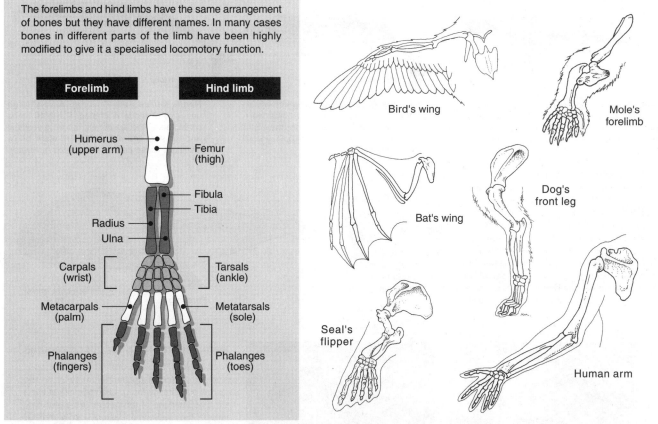

Forelimb	Hind limb

Humerus (upper arm) — Femur (thigh)
Fibula
Tibia
Radius
Ulna
Carpals (wrist) — Tarsals (ankle)
Metacarpals (palm) — Metatarsals (sole)
Phalanges (fingers) — Phalanges (toes)

Bird's wing
Mole's forelimb
Bat's wing
Dog's front leg
Seal's flipper
Human arm

1. Briefly describe the purpose of the major anatomical change that has taken place in each of the limb examples above:

   (a) Bird wing: *Highly modified for flight. Forelimb is shaped for aerodynamic lift and feather attachment.*

   (b) Human arm: _____

   (c) Seal flipper: _____

   (d) Dog foot: _____

   (e) Mole forelimb: _____

   (f) Bat wing: _____

2. Describe how **homology** in the pentadactyl limb is evidence for adaptive radiation: _____

   _____

   _____

   _____

3. Homology in the behaviour of animals (for example, sharing similar courtship or nesting rituals) is sometimes used to indicate the degree of relatedness between groups. Suggest how behaviour could be used in this way:

   _____

   _____

   _____

Vestigial Organs

Some classes of characters are more valuable than others as reliable indicators of common ancestry. Often, the less any part of an animal is used for specialised purposes, the more important it becomes for classification. This is because common ancestry is easier to detect if a particular feature is unaffected by specific adaptations arising later during the evolution of the species. Vestigial organs are an example of this because, if they have no clear function and they are no longer subject to natural selection, they will remain unchanged through a lineage. It is sometimes argued that some vestigial organs are not truly vestigial, i.e. they may perform some small function. While this may be true in some cases, the features can still be considered vestigial if their new role is a minor one, unrelated to their original function.

Ancestors of Modern Whales

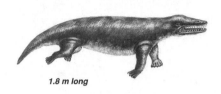

1.8 m long

2.5 m long   **20-25 m long**

Pakicetus (early Eocene) a carnivorous, four limbed, early Eocene whale ancestor, probably rather like a large otter. It was still partly terrestrial and not fully adapted for aquatic life.

Protocetus (mid Eocene). Much more whale-like than *Pakicetus*. The hind limbs were greatly reduced and although they still protruded from the body (arrowed), they were useless for swimming.

Basilosaurus (late Eocene). A very large ancestor of modern whales. The hind limbs contained all the leg bones, but were vestigial and located entirely within the main body, leaving a tissue flap on the surface (arrowed).

Vestigial organs are common in nature. The vestigial hind limbs of modern whales (right) provide anatomical evidence for their evolution from a carnivorous, four footed, terrestrial ancestor. The oldest known whale, *Pakicetus*, from the early Eocene (~54 mya) still had four limbs. By the late Eocene (~40 mya), whales were fully marine and had lost almost all traces of their former terrestrial life. For fossil evidence, see *Whale Origins* at: www.neoucom.edu/Depts/Anat/whaleorigins.htm

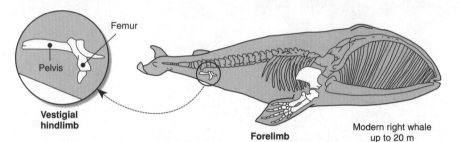

Femur

Pelvis

Vestigial hindlimb

Forelimb

Modern right whale up to 20 m

RM-DoC

Vestigial organs in birds and reptiles

In all snakes (far left), one lobe of the lung is vestigial (there is not sufficient room in the narrow body cavity for it). In some snakes there are also vestiges of the pelvic girdle and hind limbs of their walking ancestors. Like all ratites, kiwis (left) are flightless. However, more than in other ratites, the wings of kiwis are reduced to tiny vestiges. Kiwis evolved in the absence of predators to a totally ground dwelling existence.

1. In terms of natural selection explain how structures, that were once useful to an organism, could become vestigial:

_____

_____

_____

_____

2. Suggest why a vestigial structure, once it has been reduced to a certain size, may not disappear altogether:

_____

_____

3. Whale evolution shows the presence of **transitional forms** (fossils that are intermediate between modern forms and very early ancestors). Suggest how vestigial structures indicate the common ancestry of these forms:

_____

_____

_____

Biogeographical Evidence

The distribution of organisms around the world lends powerful support to the idea that modern forms evolved from ancestral populations. **Biogeography** is the study of the geographical distribution of species, both present-day and extinct. It stresses the role of dispersal of species from a point of origin across pre-existing barriers. Studies from the island populations (below) indicate that flora and fauna of different islands are more closely related to adjacent continental species than to each other.

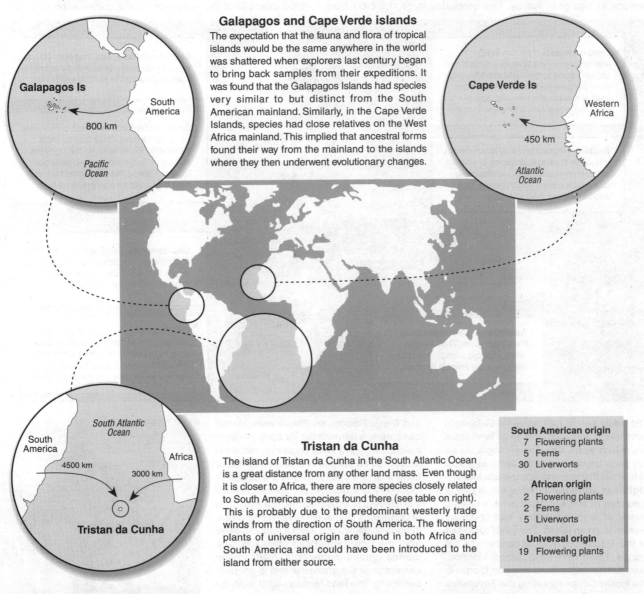

Galapagos and Cape Verde islands

The expectation that the fauna and flora of tropical islands would be the same anywhere in the world was shattered when explorers last century began to bring back samples from their expeditions. It was found that the Galapagos Islands had species very similar to but distinct from the South American mainland. Similarly, in the Cape Verde Islands, species had close relatives on the West Africa mainland. This implied that ancestral forms found their way from the mainland to the islands where they then underwent evolutionary changes.

Tristan da Cunha

The island of Tristan da Cunha in the South Atlantic Ocean is a great distance from any other land mass. Even though it is closer to Africa, there are more species closely related to South American species found there (see table on right). This is probably due to the predominant westerly trade winds from the direction of South America. The flowering plants of universal origin are found in both Africa and South America and could have been introduced to the island from either source.

South American origin
7 Flowering plants
5 Ferns
30 Liverworts

African origin
2 Flowering plants
2 Ferns
5 Liverworts

Universal origin
19 Flowering plants

1. The Galapagos Islands and the Cape Verde Islands are tropical islands close to the equator. These islands have plants and animals that are very different from each other. Explain why this is so:

_____

_____

2. The island of Tristan da Cunha is situated in the South Atlantic Ocean remote from any other land. Identify the origin of the majority of the plant species that are found there, and explain why this is so:

_____

_____

3. Using one or more specific examples, describe how **biogeography** provides support for the theory of evolution:

_____

_____

_____

_____

The Origin and Evolution of Life

Code: A 2

Oceanic Island Colonisers

Oceanic islands have a **unique biota** because only certain groups of plants and animals tend to colonise them, while others are just not able to do so. The animals that successfully colonise oceanic islands have to be marine in habit, or able to survive long periods at sea or in the air. This precludes large numbers from ever reaching distant islands. Plants also have limited capacity to reach distant islands. Only some have fruits and seeds that are salt tolerant. Many plants are transferred to the islands by wind or migrating birds. The biota of the **Galapagos islands** provide a good example of the results of such a colonisation process.

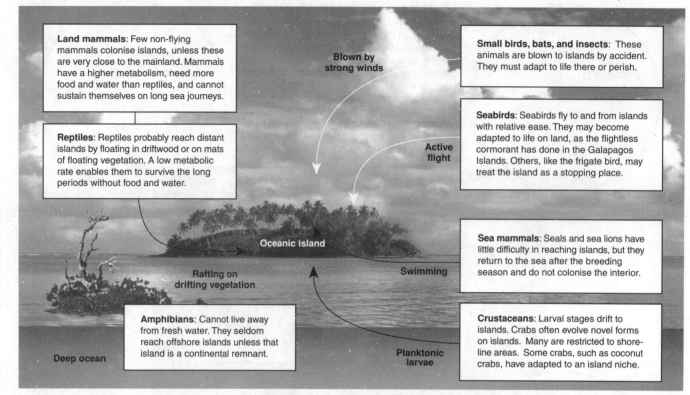

Land mammals: Few non-flying mammals colonise islands, unless these are very close to the mainland. Mammals have a higher metabolism, need more food and water than reptiles, and cannot sustain themselves on long sea journeys.

Reptiles: Reptiles probably reach distant islands by floating in driftwood or on mats of floating vegetation. A low metabolic rate enables them to survive the long periods without food and water.

Blown by strong winds

Small birds, bats, and insects: These animals are blown to islands by accident. They must adapt to life there or perish.

Seabirds: Seabirds fly to and from islands with relative ease. They may become adapted to life on land, as the flightless cormorant has done in the Galapagos Islands. Others, like the frigate bird, may treat the island as a stopping place.

Active flight

Oceanic island

Rafting on drifting vegetation

Swimming

Sea mammals: Seals and sea lions have little difficulty in reaching islands, but they return to the sea after the breeding season and do not colonise the interior.

Amphibians: Cannot live away from fresh water. They seldom reach offshore islands unless that island is a continental remnant.

Deep ocean

Planktonic larvae

Crustaceans: Larval stages drift to islands. Crabs often evolve novel forms on islands. Many are restricted to shore-line areas. Some crabs, such as coconut crabs, have adapted to an island niche.

The oldest islands making up the Galapagos archipelago appeared above sea level some 3-4 million years ago. The photographs on this page show some of the features typical of animals that colonise oceanic islands. The **flightless cormorant** (below left) is one of a number of bird species that have lost the power of flight once they had taken up residence on an island. The **giant tortoises** of the Galapagos (below centre) are not unique. There were other, almost identical, giant tortoise subspecies living on islands in the Indian Ocean including the Seychelles archipelago, Reunion, Mauritius, Farquhar, and Diego Rodriguez. These were almost completely exterminated by early Western sailors, although a small population remained untouched on the island of Aldabra. Another feature of oceanic islands is the 'adaptive radiation' (diversification) of colonising species into different specialist forms. The two forms of Galapagos iguana almost certainly arose, through diversification, from a hardy traveller from the South American mainland. The **marine iguana** (below right) feeds on the seaweeds of the shoreline and is adept at swimming. The **land iguana** (right) feeds on cacti, which are numerous.

Land iguana feeding on cactus

Flightless cormorant

Giant Galapagos tortoise

Marine iguana feeding on seaweed

1. Explain why the flora and fauna of the Galapagos Archipelago must be relatively recent arrivals:

_____

2. Describe how the marine iguana and the land iguana have become different: _____

_____

Continental Drift and Evolution

Continental drift is a measurable phenomenon; it has happened in the past and continues today. Movements of up to 2-11 cm a year have been recorded between continents using laser technology. The movements of the Earth's 12 major crustal plates are driven by thermal convection currents in the mantle; a geological process known as **plate tectonics.** Some continents appear to be drifting apart while others are on a direct collision course. Various lines of evidence show that the modern continents were once joined together as 'supercontinents'. One supercontinent, called **Gondwana**, was made up of the southern continents some 200 million years ago. The diagram below shows some of the data collected that are used as evidence to indicate how the modern continents once fitted together.

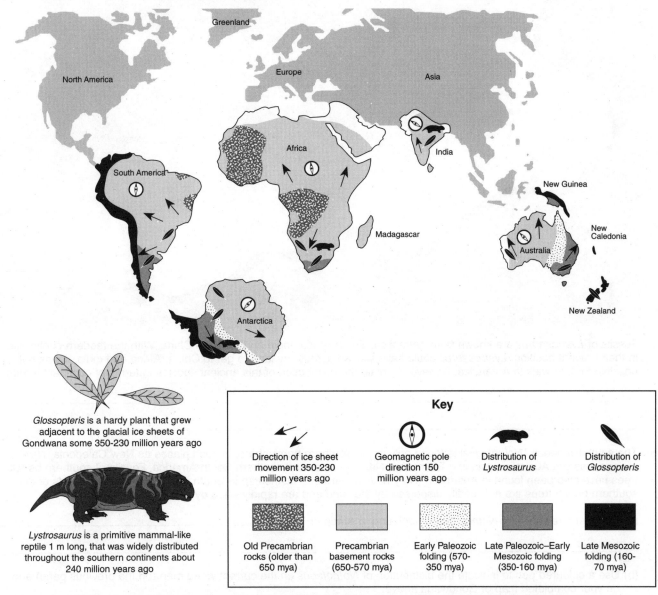

Glossopteris is a hardy plant that grew adjacent to the glacial ice sheets of Gondwana some 350-230 million years ago

Lystrosaurus is a primitive mammal-like reptile 1 m long, that was widely distributed throughout the southern continents about 240 million years ago

Key

Direction of ice sheet movement 350-230 million years ago	Geomagnetic pole direction 150 million years ago	Distribution of *Lystrosaurus*	Distribution of *Glossopteris*

Old Precambrian rocks (older than 650 mya)	Precambrian basement rocks (650-570 mya)	Early Paleozoic folding (570-350 mya)	Late Paleozoic–Early Mesozoic folding (350-160 mya)	Late Mesozoic folding (160-70 mya)

1. Name the modern landmasses (continents and large islands) that made up the supercontinent of Gondwana:

_____

2. Cut out the southern continents on page 33 and arrange them to recreate the supercontinent of Gondwana. Take care to cut the shapes out close to the coastlines. When arranging them into the space showing the outline of Gondwana on page 32, take into account the following information:
   (a) The location of ancient rocks and periods of mountain folding during different geological ages.
   (b) The direction of ancient ice sheet movements.
   (c) The geomagnetic orientation of old rocks (the way that magnetic crystals are lined up in ancient rock gives an indication of the direction the magnetic pole was at the time the rock was formed).
   (d) The distribution of fossils of ancient species such as *Lystrosaurus* and *Glossopteris*.

3. Once you have positioned the modern continents into the pattern of the supercontinent, mark on the diagram:
   (a) The likely position of the South Pole 350-230 million years ago (as indicated by the movement of the ice sheets).
   (b) The likely position of the geomagnetic South Pole 150 million years ago (as indicated by ancient geomagnetism).

4. State what general deduction you can make about the position of the polar regions with respect to land masses:

_____

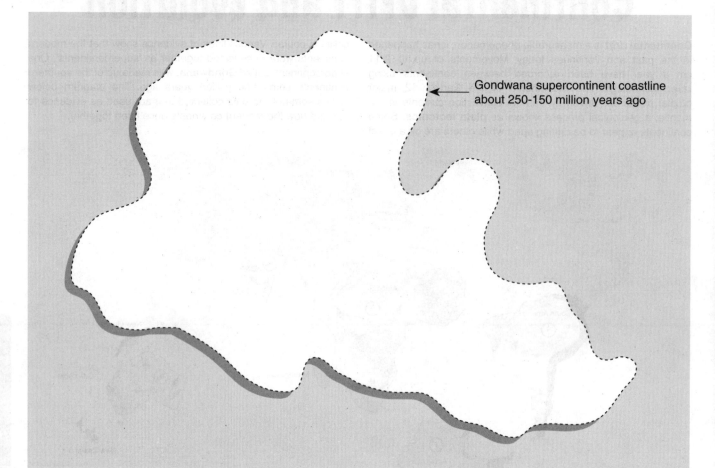

Gondwana supercontinent coastline about 250-150 million years ago

5. Fossils of *Lystrosaurus* are known from Antarctica, South Africa, India and Western China. With the modern continents in their present position, *Lystrosaurus* could have walked across dry land to get to China, Africa and India. It was not possible for it to walk to Antarctica, however. Explain the distribution of this ancient species in terms of continental drift:

_____

_____

6. The southern beech (*Nothofagus*) is found only in the southern hemisphere, in such places as New Caledonia, New Guinea, eastern Australia (including Tasmania), New Zealand, and southern South America. Fossils of southern beech trees have also been found in Antarctica. They have never been distributed in South Africa or India. The seeds of the southern beech trees are not readily dispersed by the wind and are rapidly killed by exposure to salt water.

(a) Suggest a reason why *Nothofagus* is not found in Africa or India: _____

_____

(b) Use a coloured pen to indicate the distribution of *Nothofagus* on the current world map (on the previous page) and on your completed map of Gondwana above.

(c) State how the arrangement of the continents into Gondwana explains this distribution pattern:

_____

_____

7. The Atlantic Ocean is currently opening up at the rate of 2 cm per year. At this rate in the past, calculate how long it would have taken to reach its current extent, with the distance from Africa to South America being 2300 km (assume the rate of spreading has been constant):

_____

8. Explain how continental drift provides evidence to support evolutionary theory: _____

_____

_____

_____

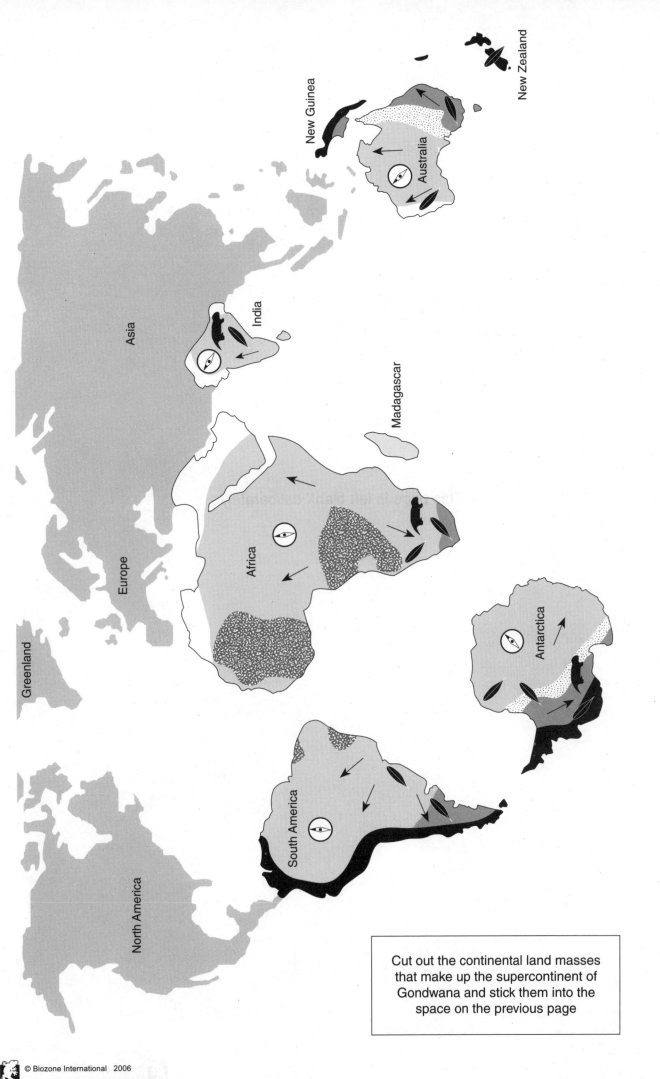

New Guinea

New Zealand

Australia

India

Asia

Madagascar

Europe

Africa

Greenland

Antarctica

North America

South America

Cut out the continental land masses
that make up the supercontinent of
Gondwana and stick them into the
space on the previous page

This page is left blank deliberately

Mechanisms of Evolution

The modern synthesis of evolution and the mechanisms of evolution

The modern theory of evolution, gene pools and evolution, the species concept, species life cycle, allopatric and sympatric speciation

Learning Objectives

☐ 1. Compile your own glossary from the KEY WORDS displayed in **bold type** in the learning objectives below.

Modern Synthesis of Evolution *(pages 11, 38-40)*

☐ 2. Define the term **evolution**, explaining how evolution is a feature of **populations** and not of individuals.

☐ 3. Identify some of the main contributors to the modern theory of evolution. Include reference to:
(a) Lamarck's theory of evolution by the inheritance of acquired characteristics and its problems.
(b) Darwin's theory of evolution by natural selection.

☐ 4. Discuss the evidence for various theories for the origin of species, including the theories of Darwin and Wallace, and Lamarck, and the theories of **panspermia** and evolution by special creation. Discuss the applicability of the scientific method to each.

☐ 5. Outline the fundamental ideas in Darwin's theory of evolution by natural selection. Include reference to:
• The tendency of populations to overproduce.
• The fact that overproduction leads to competition.
• The fact that members of a species show variation, that sexual reproduction promotes variation, and that variation is (usually) heritable.
• The differential survival and reproduction of individuals with favourable, heritable variations.

☐ 6. Appreciate how Darwin's original theory has been modified in the **new synthesis** to incorporate our current understanding of genetics and inheritance.

☐ 7. Understand the term **fitness** and explain how evolution, through **adaptation**, equips species for survival. Recognise structural and physiological adaptations of organisms to their environment.

The Species Concept *(pages 65-66)*

☐ 8. Explain what is meant by the term, **species**. Giving examples, describe how the nature of some species creates problems for our definition. Distinguish between **species**, **sub species**, and **hybrids**.

☐ 9. Describe the concept of a **ring species** and closely related species where the reproductive isolation is no longer maintained. Provide examples of these species.

Concept of the Gene Pool *(pages 37, 49-57)*

☐ 10. Understand the concept of the **gene pool** and explain the term **deme**. Recognise that populations may be of various sizes and geographical extent.

☐ 11. Explain the term **allele frequency** and describe how allele frequencies are expressed for a population.

☐ 12. State the Hardy-Weinberg principle (of **genetic equilibrium**). Understand the criteria that must be satisfied in order to achieve genetic equilibrium in a population. Identify the consequences of the fact that these criteria are rarely met in reality.

☐ 13. Explain how the **Hardy-Weinberg equation** provides a simple mathematical model of genetic equilibrium in a population. Demonstrate an ability to use the Hardy-Weinberg equation to calculate the allele, genotype, and phenotype frequencies from appropriate data.

Microevolution *(pages 40-50, 57-60, 72)*

☐ 14. Recognise that changes occur in gene pools (**microevolution**) when any or all of the criteria for genetic equilibrium are not met. Appreciate that it is populations, not individuals, that evolve.

☐ 15. Recognise the forces in microevolution that may alter allele frequencies: **natural selection**, **genetic drift**, **gene flow**, and **mutation**. Identify processes that increase genetic variation and those that decrease it.

☐ 16. Recall that heritable variation is the raw material for natural selection. Explain how **natural selection** is responsible for most evolutionary change by selectively changing genetic variation through differential survival and reproduction. Interpret data to explain how natural selection produces change in a population (see #17).

☐ 17. Describe three types of natural selection: **stabilising**, **directional**, and **disruptive selection**. Describe the outcome of each type in a population exhibiting a normal curve in phenotypic variation.

☐ 18. As required, describe examples of evolution including:
(a) **Transient polymorphism**, e.g. **industrial melanism** in peppered moths (*Biston betularia*).
(b) The sickle cell trait as the basis for **balanced polymorphism** in malarial regions.
(c) Changes to the size and shape of the beaks of **Galapagos finches**.
(d) The development of **multiple antibiotic resistance** in bacteria as an example of rapid evolution in response to environmental change.

☐ 19. Define the term **genetic drift** and describe the conditions under which it is important. Explain, using diagrams or a gene pool model, how genetic drift may lead to loss or **fixation of alleles** (where a gene is represented in the population by only one allele).

☐ 20. Recognise **mutations** as the source of all new alleles. Use a diagram to explain how mutations alter the genetic equilibrium of a population. Recall that recombination during meiosis reshuffles alleles and increases variation, but it does not create new alleles.

□ 21. Explain how **migration** leads to **gene flow** between natural populations, and may affect allele frequencies.

□ 22. Explain each of the following special events in gene pools. In each case, include reference to the genetic consequences and explain how each of these may accelerate the pace of evolutionary change.
(a) The **founder effect**
(b) The **population bottleneck effect**

□ 23. Appreciate how the founder effect and population (genetic) bottlenecks may accelerate the pace of evolutionary change. Explain the importance of **genetic drift** in populations that undergo these events.

Sexual Selection *(page 46)*

□ 24. Discuss the role of **sexual selection** in affecting anatomy and behaviour. Suggest how sexual selection may lead to the evolution of elaborate secondary sexual characteristics, particularly in males.

Speciation *(pages 63-74)*

□ 25. Recognise the role of **natural selection** and **isolation** in **speciation**. Discuss speciation in terms of migration, geographical or ecological isolation and adaptation, leading to the reproductive or genetic isolation of gene pools. Distinguish between the **allopatric** and **sympatric** distribution of populations.

□ 26. Describe the mechanisms through which populations achieve or maintain **reproductive isolation** including: geographical isolation, polyploidy, changes to behaviour, morphology, or physiology, and niche differentiation and **character displacement**. If required, distinguish between and describe **prezygotic** and **postzygotic** reproductive isolating mechanisms.

□ 27. Explain the events occurring in **allopatric speciation**, identifying situations in which it is most likely to occur.

□ 28. Explain the events occurring in **sympatric speciation** and describe the situations in which it is most likely to occur. Explain why reproductive isolating mechanisms tend to be much more pronounced between sympatric (as opposed to allopatric) species. Discuss the role of **polyploidy** in instant speciation events.

□ 29. Recognise stages in species development, including reference to the reduction in gene flow as populations become increasingly isolated.

□ 30. Describe the major stages in a **species life cycle** extending from origin to extinction.

Artificial Selection *(pages 61-64)*

□ 31. Explain the genetic basis of **artificial selection** (**selective breeding**). Describe examples of artificial selection, explaining how the selection process has led to the development of particular traits:
• Crop plants, e.g. brassicas, wheat, maize.
• Companion animals (e.g. dogs).

□ 32. Using examples, explain the terms **outbreeding**, **inbreeding**, interspecific **hybridisation**, and **polyploidy**. Using examples, explain what is meant by F_1 **hybrid vigour** and explain its genetic basis.

See page 7 for additional details of these texts:
■ Clegg, C.J., 1999. **Genetics and Evolution**, (John Murray), pp. 60-78.

■ Futuyma, D.J., 2005. **Evolution**, (Sinauer Associates), chpt. 9-17 as required.

■ Helms, D.R. *et al.*, 1998. **Biology in the Laboratory** (W.H. Freeman), #20, #21.

■ Jones, N., *et al.*, 2001. **The Essentials of Genetics**, pp. 190-232.

■ Martin, R.A., 2004. **Missing Links**, chpt. 2 and case histories from section II as required.

■ Zimmer, C., 2001. **Evolution: The Triumph of an Idea**, (HarperCollins), chpt. 1-4, 6 as required.

See page 7 for details of publishers of periodicals:

STUDENT'S REFERENCE

■ **The Hardy-Weinberg Principle** Biol. Sci. Rev., 15(4), April 2003, pp. 7-9. *A succinct explanation of the basis of the Hardy-Weinberg principle, and its uses in estimating genotype frequencies and predicting change in populations.*

■ **Speciation** Biol. Sci. Rev., 16(2) Nov. 2003, pp. 24-28. *An excellent account of speciation. Case examples include the cichlids of Lake Victoria and the founder effect in mynahs.*

■ **Optimality** Biol. Sci. Rev., 17(4), April 2005, pp. 2-5. *Environmental stability and optimality of structure and function can explain evolutionary stasis in animals. Examples are described.*

■ **Polymorphism** Biol. Sci. Rev., 14(1), Sept. 2001, pp. 19-21. *An account of polymorphism in populations, with several case studies (including Biston moths) provided as illustrative examples.*

■ **The Species Enigma** New Scientist, 13 June 1998 (Inside Science). *The nature of species, ring species, and the status of hybrids.*

■ **Together We're Stronger** New Scientist, 15 March 2003. (Inside Science). *The mechanisms behind the evolution of social behaviour in animals. The evolution of eusociality in hymenopteran insects is the case study provided.*

■ **Listen, We're Different** New Scientist, 17 July 1999, pp. 32-35. *An excellent account of speciation in periodic cicadas as a result of behavioural and temporal isolating mechanisms.*

■ **Plants on the Move** New Scientist, 20 March 1999 (Inside Science). *Glaciation and warming have evolutionary consequences for flora.*

TEACHER'S REFERENCE

■ **Evolution: Five Big Questions** New Scientist, 14 June 2003, pp. 32-39, 48-51. *A discussion of the most covered points regarding evolution and the mechanisms by which it occurs.*

■ **15 Answers to Creationist Nonsense** Scientific American, July 2002, pp. 62-69. *A synopsis of the common arguments presented by Creationists and the answers offered by science.*

■ **Cichlids of the Rift Lakes** Scientific American, Feb. 1999, pp. 44-49. *An excellent account of the speciation events documented in cichlid fishes.*

■ **Fair Enough** New Scientist, 12 October 2002, pp. 34-37. *The inheritance of skin colour in humans. This article examines why humans have such varied skin pigmentation and looks at the argument for there being a selective benefit to being dark or pale in different environments.*

■ **Skin Deep** Scientific American, October 2002, pp. 50-57. *This article examines the evolution of skin colour in humans and presents powerful evidence for skin colour ("race") being the end result of opposing selection forces (the need for protection of folate from UV vs the need to absorb vitamin D). Clearly written and of high interest, this is a must for student discussion and a perfect vehicle for examining natural selection.*

■ **Replaying Life** New Scientist, 13 February 1999, pp. 29-33. *Rapid evolution in bacteria driven by habitat diversity and niche differentiation.*

■ **How the Species Became** New Scientist, 11 Oct. 2003, pp. 32-35. *Stability in species and new ideas on speciation. Species are stable if changes in form or behaviour are damped, but unstable if the changes escalate as new generations shuffle parental genes and natural selection discards the allele combinations that do not work well.*

■ **Live and Let Live** New Scientist, 3 July 1999, pp. 32-36. *Recent research suggests that hybrids are intact entities subject to the same evolutionary pressures as pure species.*

See pages 4-5 for details of how to access **Bio Links** from our web site: **www.thebiozone.com** From Bio Links, access sites under the topics:
GENERAL BIOLOGY ONLINE RESOURCES > Online Textbooks and Lecture Notes: • An on-line biology book... *and others* > **General Online Biology resources:** • Ken's Bioweb resources ... *and others* > **Glossaries:** • Evolutionary biology and genetics glossary... *and others*

EVOLUTION: • Enter evolution: theory & history • BIO 414 evolution • Evolution • Evolution on the web for biology students • Harvard University biology links: evolution • The Talk.Origins archive • speciation ... *and others* > **Charles Darwin:** • Darwin and evolution overview ... *and others*

GENETICS > Population Genetics: • Industrial melanism in *Biston betularia* • Introduction to evolutionary biology • Microevolution and population genetics • Random genetic drift • Population genetics: lecture notes ... *and others*

Presentation MEDIA to support this topic:

EVOLUTION

Evolution

Genes and Evolution

Each individual in a population is the carrier of its own particular combination of genetic material. Different combinations of genes come about because of the shuffling of the chromosomes during gamete formation. New combinations of alleles arise as a result of mate selection and the chance meeting of a vast range of different gametes from each of the two parents. Some combinations are well suited to particular environments, while others are not. Those organisms with an inferior collection of genes will have reduced reproductive success. This means that the genes (alleles) they carry will decrease in frequency and fewer will be passed on to the next generation's gene pool. Those individuals with more successful allele combinations will have higher reproductive success. The frequency of their alleles in the gene pool will increase.

The Importance of Genetic Processes in Evolution

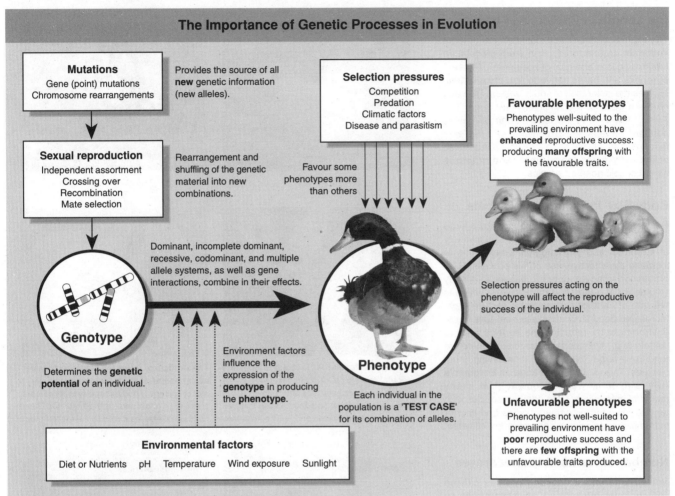

Mutations
Gene (point) mutations
Chromosome rearrangements

Provides the source of all **new** genetic information (new alleles).

Sexual reproduction
Independent assortment
Crossing over
Recombination
Mate selection

Rearrangement and shuffling of the genetic material into new combinations.

Selection pressures
Competition
Predation
Climatic factors
Disease and parasitism

Favour some phenotypes more than others

Favourable phenotypes
Phenotypes well-suited to the prevailing environment have **enhanced** reproductive success: producing **many offspring** with the favourable traits.

Dominant, incomplete dominant, recessive, codominant, and multiple allele systems, as well as gene interactions, combine in their effects.

Genotype

Determines the **genetic potential** of an individual.

Environment factors influence the expression of the **genotype** in producing the **phenotype**.

Phenotype

Each individual in the population is a '**TEST CASE**' for its combination of alleles.

Selection pressures acting on the phenotype will affect the reproductive success of the individual.

Unfavourable phenotypes
Phenotypes not well-suited to prevailing environment have **poor** reproductive success and there are **few offspring** with the unfavourable traits produced.

Environmental factors

Diet or Nutrients   pH   Temperature   Wind exposure   Sunlight

1. Discuss the role of sexual reproduction and selection in evolution:

_____

_____

_____

_____

_____

_____

2. Describe the long-term effect on the gene pool of enhanced reproductive success for a particular phenotype:

_____

_____

3. Explain why each individual in a population is a **test case** for its combination of alleles: _____

_____

_____

_____

Mechanisms of Evolution

Code: A 3

Adaptations and Fitness

An **adaptation**, is any heritable trait that suits an organism to its natural function in the environment (its niche). These traits may be structural, physiological, or behavioural. The idea is important for evolutionary theory because adaptive features promote fitness. **Fitness** is a measure of an organism's ability to maximise the numbers of offspring surviving to reproductive age. Adaptations are distinct from properties which, although they may be striking, cannot be described as adaptive unless they are shown to be functional in the organism's natural habitat. Genetic adaptation must not be confused with **physiological adjustment** (acclimatisation), which refers to an organism's ability to adapt during its lifetime to changing environmental conditions (e.g. a person's acclimatisation to altitude). Examples of adaptive features arising through evolution are illustrated below.

Ear Length in Rabbits and Hares

The external ears of many mammals are used as important organs to assist in thermoregulation (controlling loss and gain of body heat). The ears of rabbits and hares native to hot, dry climates, such as the jack rabbit of south-western USA and northern Mexico, are relatively very large. The Arctic hare lives in the tundra zone of Alaska, northern Canada and Greenland, and has ears that are relatively short. This reduction in the size of the extremities (ears, limbs, and noses) is typical of cold adapted species.

Arctic hare: *Lepus arcticus*

Black-tail jackrabbit: *Lepus californicus*

Body Size in Relation to Climate

Regulation of body temperature requires a large amount of energy and mammals exhibit a variety of structural and physiological adaptations to increase the effectiveness of this process. Heat production in any endotherm depends on body volume (heat generating metabolism), whereas the rate of heat loss depends on surface area. Increasing body size minimises heat loss to the environment by reducing the surface area to volume ratio. Animals in colder regions therefore tend to be larger overall than those living in hot climates. This relationship is know as **Bergman's rule** and it is well documented in many mammalian species. Cold adapted species also tend to have more compact bodies and shorter extremities than related species in hot climates.

Fennec fox

Arctic fox

The **fennec fox** of the Sahara illustrates the adaptations typical of mammals living in hot climates: a small body size and lightweight fur, and long ears, legs, and nose. These features facilitate heat dissipation and reduce heat gain.

The **Arctic fox** shows the physical characteristics typical of cold-adapted mammals: a stocky, compact body shape with small ears, short legs and nose, and dense fur. These features reduce heat loss to the environment.

Number of Horns in Rhinoceroses

Not all differences between species can be convincingly interpreted as adaptations to particular environments. Rhinoceroses charge rival males and predators, and the horn(s), when combined with the head-down posture, add effectiveness to this behaviour. Horns are obviously adaptive, but it is not clear that the possession of one (Indian rhino) or two (black rhino) horns is necessarily related directly to the environment in which those animals live.

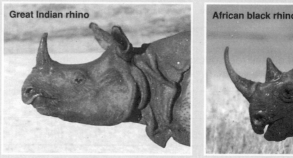
Great Indian rhino

African black rhino

1. Distinguish between adaptive features (genetic) and acclimatisation: _____

_____

_____

2. Explain the nature of the relationship between the length of extremities (such as limbs and ears) and climate: _____

_____

_____

3. Explain the adaptive value of a larger body size at high latitude: _____

_____

_____

The Modern Theory of Evolution

Although **Charles Darwin** is credited with the development of the theory of evolution by natural selection, there were many people that contributed ideas upon which he built his own. Since Darwin first proposed his theory, aspects that were problematic (such as the mechanism of inheritance) have now been explained. The development of the modern theory of evolution has a history going back at least two centuries. The diagram below illustrates the way in which some of the major contributors helped to form the currently accepted model, or **new synthesis**. Understanding of evolutionary processes continued to grow through the 1980s and 1990s as comparative molecular sequence data were amassed and understanding of the molecular basis of developmental mechanisms improved. Most recently, in the exciting new area of evolutionary developmental biology (**evo-devo**), biologists have been exploring how developmental gene expression patterns explain how groups of organisms evolved.

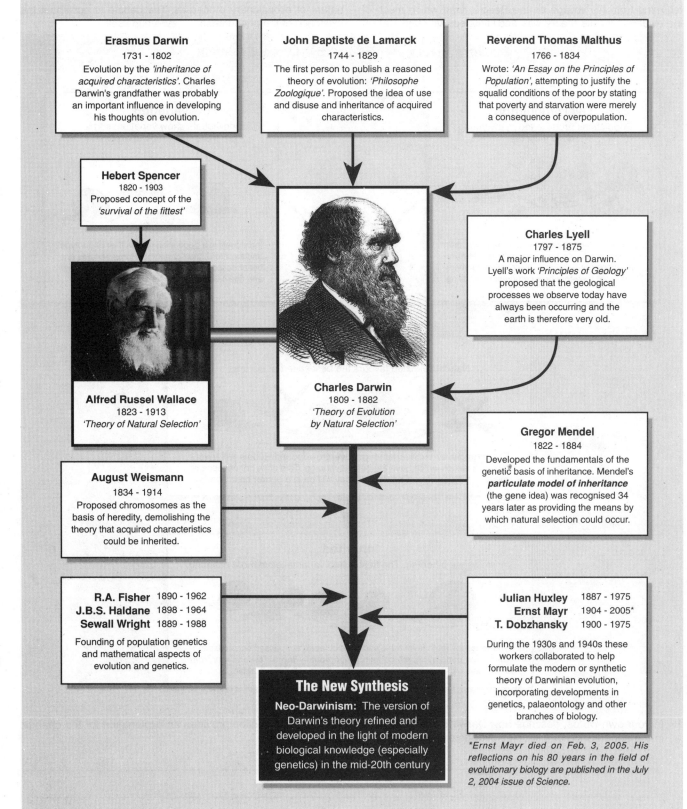

Erasmus Darwin
1731 - 1802
Evolution by the *'inheritance of acquired characteristics'*. Charles Darwin's grandfather was probably an important influence in developing his thoughts on evolution.

John Baptiste de Lamarck
1744 - 1829
The first person to publish a reasoned theory of evolution: *'Philosophe Zoologique'*. Proposed the idea of use and disuse and inheritance of acquired characteristics.

Reverend Thomas Malthus
1766 - 1834
Wrote: *'An Essay on the Principles of Population'*, attempting to justify the squalid conditions of the poor by stating that poverty and starvation were merely a consequence of overpopulation.

Hebert Spencer
1820 - 1903
Proposed concept of the *'survival of the fittest'*

Charles Lyell
1797 - 1875
A major influence on Darwin. Lyell's work *'Principles of Geology'* proposed that the geological processes we observe today have always been occurring and the earth is therefore very old.

Alfred Russel Wallace
1823 - 1913
'Theory of Natural Selection'

Charles Darwin
1809 - 1882
'Theory of Evolution by Natural Selection'

Gregor Mendel
1822 - 1884
Developed the fundamentals of the genetic basis of inheritance. Mendel's **particulate model of inheritance** (the gene idea) was recognised 34 years later as providing the means by which natural selection could occur.

August Weismann
1834 - 1914
Proposed chromosomes as the basis of heredity, demolishing the theory that acquired characteristics could be inherited.

R.A. Fisher 1890 - 1962
J.B.S. Haldane 1898 - 1964
Sewall Wright 1889 - 1988
Founding of population genetics and mathematical aspects of evolution and genetics.

Julian Huxley 1887 - 1975
Ernst Mayr 1904 - 2005*
T. Dobzhansky 1900 - 1975
During the 1930s and 1940s these workers collaborated to help formulate the modern or synthetic theory of Darwinian evolution, incorporating developments in genetics, palaeontology and other branches of biology.

The New Synthesis
Neo-Darwinism: The version of Darwin's theory refined and developed in the light of modern biological knowledge (especially genetics) in the mid-20th century

Ernst Mayr died on Feb. 3, 2005. His reflections on his 80 years in the field of evolutionary biology are published in the July 2, 2004 issue of Science.

Mechanisms of Evolution

1. From the diagram above, choose one of the contributors to the development of evolutionary theory (excluding Charles Darwin himself), and write a few paragraphs discussing their role in contributing to Darwin's ideas. You may need to consult an encyclopaedia or other reference to assist you.

Code: RA 3

Darwin's Theory

In 1859, Darwin and Wallace jointly proposed that new species could develop by a process of natural selection. Natural selection is the term given to the mechanism by which better adapted organisms survive to produce a greater number of viable offspring. This has the effect of increasing their proportion in the population so that they become more common. It is Darwin who is best remembered for the theory of evolution by natural selection through his famous book: **'On the origin of species by means of natural selection'**, written 23 years after returning from his voyage on the Beagle, from which much of the evidence for his theory was accumulated. Although Darwin could not explain the origin of variation nor the mechanism of its transmission (this was provided later by Mendel's work), his basic theory of evolution by natural selection (outlined below) is widely accepted today. The study of population genetics has greatly improved our understanding of evolutionary processes, which are now seen largely as a (frequently gradual) change in allele frequencies within a population. Students should be aware that scientific debate on the subject of evolution centres around the relative merits of various alternative hypotheses about the nature of evolutionary processes. The debate is not about the existence of the phenomenon of evolution itself.

Darwin's Theory of Evolution by Natural Selection

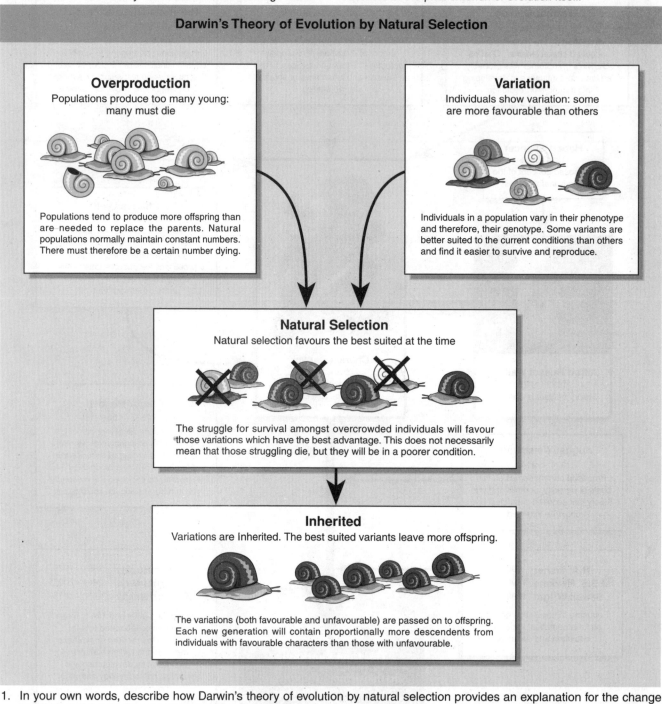

Overproduction
Populations produce too many young: many must die

Populations tend to produce more offspring than are needed to replace the parents. Natural populations normally maintain constant numbers. There must therefore be a certain number dying.

Variation
Individuals show variation: some are more favourable than others

Individuals in a population vary in their phenotype and therefore, their genotype. Some variants are better suited to the current conditions than others and find it easier to survive and reproduce.

Natural Selection
Natural selection favours the best suited at the time

The struggle for survival amongst overcrowded individuals will favour those variations which have the best advantage. This does not necessarily mean that those struggling die, but they will be in a poorer condition.

Inherited
Variations are Inherited. The best suited variants leave more offspring.

The variations (both favourable and unfavourable) are passed on to offspring. Each new generation will contain proportionally more descendents from individuals with favourable characters than those with unfavourable.

1. In your own words, describe how Darwin's theory of evolution by natural selection provides an explanation for the change in the appearance of a species over time:

_____

_____

_____

_____

Natural Selection

Natural selection operates on the phenotypes of individuals, produced by their particular combinations of alleles. In natural populations, the allele combinations of some individuals are perpetuated at the expense of other genotypes. This differential survival of some genotypes over others is called **natural selection**. The effect of natural selection can vary; it can act to maintain the genotype of a species or to change it.

Stabilising selection maintains the established favourable characteristics and is associated with stable environments. In contrast, **directional selection** favours phenotypes at one extreme of the phenotypic range and is associated with gradually changing environments. **Disruptive selection** is a much rarer form of selection favouring two phenotypic extremes, and is a feature of fluctuating environments.

Stabilising Selection

Extreme variations are culled from the population (there is selection against them). Those with the established (middle range) adaptive phenotype are retained in greater numbers. This reduces the variation for the phenotypic character. In the example right, light and dark snails are eliminated, leaving medium coloured snails. Stabilising selection can be seen in the selection pressures on human birth weights.

Directional Selection

Directional selection is associated with gradually changing conditions, where the adaptive phenotype is shifted in one direction and one aspect of a trait becomes emphasised (e.g. colouration). In the example right, light coloured snails are eliminated and the population becomes darker. Directional selection was observed in peppered moths in England during the Industrial Revolution. They responded to the air pollution of industrialisation by increasing the frequency of darker, melanic forms.

Disruptive or Diversifying Selection

Disruptive selection favours two extremes of a trait at the expense of intermediate forms. It is associated with a fluctuating environment and gives rise to **balanced polymorphism** in the population. In the example right, there is selection against medium coloured snails, which are eliminated. There is considerable evidence that predators, such as insectivorous birds, are more likely to find and eat common morphs and ignore rare morphs. This enables the rarer forms to persist in the population.

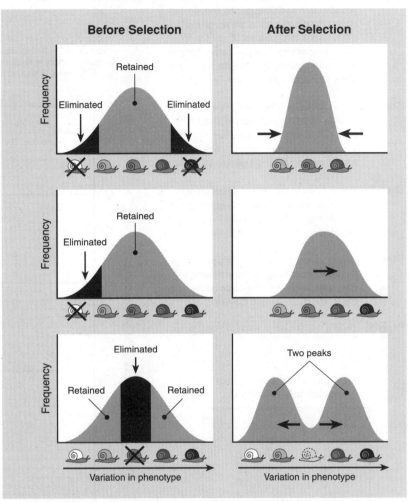

1. (a) Distinguish between **directional selection** and **disruptive selection**, identifying when each is likely to operate:

   _____

   _____

   _____

   (b) State which of the three types of selection described above will lead to evolution, and explain why:

   _____

   _____

2. Explain how a change in environment may result in selection becoming directional rather than stabilising:

   _____

   _____

3. Explain how, in a population of snails, through natural selection, shell colour could change from light to dark over time:

   _____

   _____

   _____

Mechanisms of Evolution

Code: A 3

Selection for Human Birth Weight

Selection pressures operate on populations in such a way as to reduce mortality. For humans, giving birth is a special, but often traumatic, event. In a study of human birth weights it is possible to observe the effect of selection pressures operating to constrain human birth weight within certain limits. This is a good example of **stabilising selection**. This activity explores the selection pressures acting on the birth weight of human babies. Carry out the steps below:

Step 1: Collect the birth weights from 100 birth notices from your local newspaper (or 50 if you are having difficulty getting enough; this should involve looking back through the last 2-3 weeks of birth notices). If you cannot obtain birth weights in your local newspaper, a set of 100 sample birth weights is provided in the Model Answers booklet.

Step 2: Group the weights into each of the 12 weight classes (of 0.5 kg increments). Determine what percentage (of the total sample) fall into each weight class (e.g. 17 babies weigh 2.5-3.0 kg out of the 100 sampled = 17%)

Step 3: Graph these in the form of a histogram for the 12 weight classes (use the graphing grid provided right). Be sure to use the scale provided on the left vertical (y) axis.

Step 4: Create a second graph by plotting percentage mortality of newborn babies in relation to their birth weight. Use the scale on the right y axis and data provided (below).

Step 5: Draw a line of 'best fit' through these points.

Mortality of newborn babies related to birth weight

Weight (kg)	Mortality (%)
1.0	80
1.5	30
2.0	12
2.5	4
3.0	3
3.5	2
4.0	3
4.5	7
5.0	15

Source: Biology: The Unity & Diversity of Life (4th ed), by Starr and Taggart

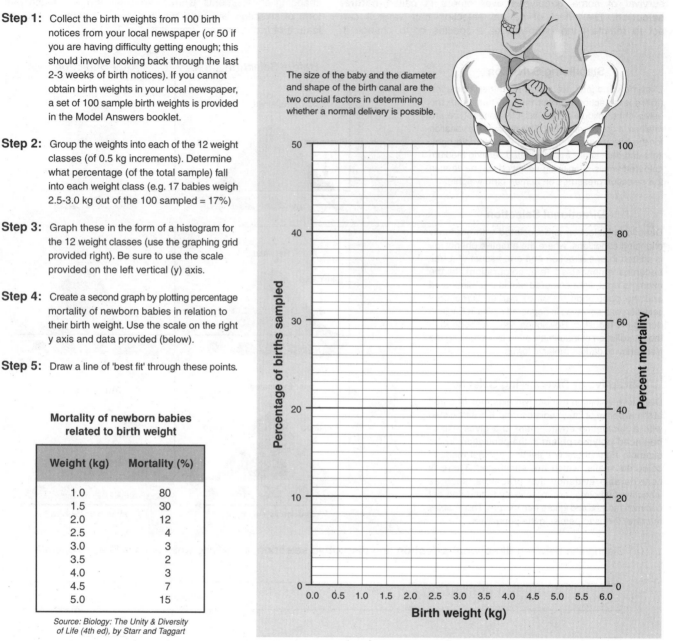

The size of the baby and the diameter and shape of the birth canal are the two crucial factors in determining whether a normal delivery is possible.

1. Describe the shape of the histogram for birth weights: _____

2. State the optimum birth weight in terms of the lowest newborn mortality: _____

3. Describe the relationship between the newborn mortality and the birth weights: _____

_____

4. Describe the selection pressures that are operating to control the range of birth weight: _____

_____

_____

5. Describe how medical intervention methods during pregnancy and childbirth may have altered these selection pressures:

_____

_____

Industrial Melanism

Natural selection may act on the frequencies of phenotypes (and hence genotypes) in populations in one of three different ways (through stabilising, directional, or disruptive selection). Over time, natural selection may lead to a permanent change in the genetic makeup of a population. The increased prevalance of melanic forms of the peppered moth, *Biston betularia*, during the Industrial Revolution is one of the best known examples of directional selection following a change in environmental conditions. Although the protocols used in the central experiments on *Biston*, and the conclusions drawn from these experiments, have been called into question, the collections of moths do provide documented evidence of phenotypic change.

Industrial melanism in peppered moths, *Biston betularia*

The **peppered moth**, *Biston betularia*, occurs in two forms (morphs): the grey mottled form, and a dark melanic form. Changes in the relative abundance of these two forms was hypothesised to be the result of selective predation by birds, with pale forms suffering higher mortality in industrial areas because they are more visible. The results of experiments by H.D. Kettlewell supported this hypothesis but did not confirm it, since selective predation by birds was observed but not quantified. Other research indicates that predation by birds is not the only factor determining the relative abundance of the different colour morphs.

Grey or mottled morph: vulnerable to predation in industrial areas where the trees are dark.

Melanic or carbonaria morph: dark colour makes it less vulnerable to predation in industrial areas.

Museum collections of the peppered moth made over the last 150 years show a marked change in the frequency of the melanic form. Moths collected in 1850 (above left), prior to the major onset of the industrial revolution in England. Fifty years later (above right) the frequency of the darker melanic forms had greatly increased. Even as late as the mid 20th century, coal-based industries predominated in some centres, and the melanic form occurred in greater frequency in these areas (see map, right).

Frequency of peppered moth forms in 1950

This map shows the relative frequencies of the two forms of peppered moth in the UK in 1950; a time when coal-based industries still predominated in some major centres.

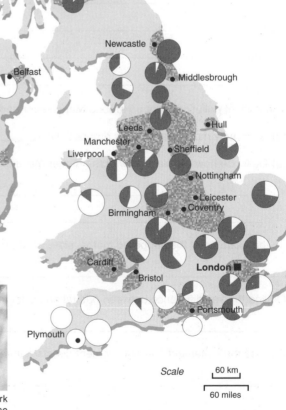

Scale 60 km
60 miles

Key to frequency graphs

Grey or speckled form

Melanic or carbonaria form

Industrial areas

Non-industrial areas

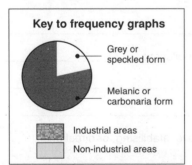

A grey (mottled) form of *Biston*, camouflaged against a lichen covered bark surface. In the absence of soot pollution, mottled forms appear to have the selective advantage.

A melanic form of *Biston*, resting on a dark branch, so that it appears as part of the branch. Note that the background has been faded out so that the moth can be seen.

Mechanisms of Evolution

Code: RDA 2

Changes in frequency of melanic peppered moths

In the 1940s and 1950s, coal burning was still at intense levels around the industrial centres of Manchester and Liverpool. During this time, the melanic form of the moth was still very dominant. In the rural areas further south and west of these industrial centres, the grey or speckled forms increased dramatically. With the decline of coal burning factories and the Clean Air Acts in cities, the air quality improved between 1960 and 1980. Sulfur dioxide and smoke levels dropped to a fraction of their previous levels. This coincided with a sharp fall in the relative numbers of melanic moths.

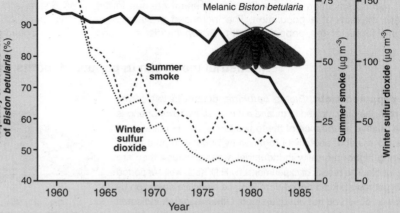

Frequency of melanic peppered moth related to reduced air pollution

1. The populations of peppered moth in England have undergone changes in the frequency of an obvious phenotypic character over the last 150 years. Describe the phenotypic character that changed in its frequency:

   _____

2. (a) Identify the (proposed) selective agent for phenotypic change in *Biston*: _____

   (b) Describe how the selection pressure on the light coloured morph has changed with changing environmental conditions over the last 150 years:

   _____

   _____

3. The industrial centres for England in 1950 were located around London, Birmingham, Liverpool, Manchester, and Leeds. Glasgow in Scotland also had a large industrial base. Comment on how the relative frequencies of the two forms of peppered moth were affected by the geographic location of industrial regions:

   _____

   _____

   _____

   _____

4. The level of pollution dropped around Manchester and Liverpool between 1960 and 1985.

   (a) State how much the pollution dropped by: _____

   (b) Describe how the frequency of the darker melanic form responded to this reduced pollution: _____

   _____

   _____

5. In the example of the peppered moths, state whether the selection pressure is disruptive, stabilising, or directional:

   _____

6. Outline the key difference between natural and artificial selection: _____

   _____

7. Discuss the statement "the environment directs natural selection": _____

   _____

   _____

   _____

Heterozygous Advantage

There are two mechanisms by which natural selection can affect allele frequencies. Firstly, there may be selection against one of the homozygotes. When one homozygous type (for example, aa), has a lower fitness than the other two genotypes (in this case, Aa or AA), the frequency of the deleterious allele will tend to decrease until it is completely eliminated. In some situations, both homozygous conditions (aa and AA) have lower fitness than the heterozygote; a situation that leads to **heterozygous advantage** and may result in the stable coexistence of both alleles in the population (**balanced polymorphism**). There are remarkably few well-documented examples in which the evidence for heterozygous advantage is conclusive. The maintenance of the sickle cell mutation in malaria-prone regions is one such example.

The Sickle Cell Allele (Hbˢ)

Sickle cell disease is caused by a mutation to a gene that directs the production of the human blood protein called haemoglobin. The mutant allele is known as **Hbˢ** and produces a form of haemoglobin that differs from the normal form by just one amino acid in the β-chain. This minute change however causes a cascade of physiological problems in people with the allele. Some of the red blood cells containing mutated haemoglobin alter their shape to become irregular and spiky; the so-called **sickle cells**.

Sickle cells have a tendency to clump together and work less efficiently. In people with just one sickle cell allele plus a normal allele (the heterozygote condition **HbˢHb**), there is a mixture of both red blood cell types and they are said to have the sickle cell trait. They are generally unaffected by the disease except in low oxygen environments (e.g. climbing at altitude). People with two Hbˢ genes (**HbˢHbˢ**) suffer severe illness and even death. For this reason Hbˢ is considered **a lethal gene**.

Heterozygous Advantage in Malarial Regions

Falciparum malaria is widely distributed throughout central Africa, the Mediterranean, Middle East, and tropical and semi-tropical Asia (Fig. 1). It is transmitted by the *Anopheles* mosquito, which spreads the protozoan *Plasmodium falciparum* from person to person as it feeds on blood.

SYMPTOMS: These appear 1-2 weeks after being bitten, and include headache, shaking, chills, and fever. Falciparum malaria is more severe than other forms of malaria, with high fever, convulsions, and coma. It can be fatal within days of the first symptoms appearing.

THE PARADOX: The Hbˢ allele offers considerable protection against malaria. Sickle cells have low potassium levels, which causes plasmodium parasites inside these cells to die. Those with a normal phenotype are very susceptible to malaria, but heterozygotes (**HbˢHb**) are much less so. This situation, called **heterozygous advantage**, has resulted in the Hbˢ allele being present in moderately high frequencies in parts of Africa and Asia despite its harmful effects (Fig. 2). This is a special case of balanced polymorphism, called a **balanced lethal system** because neither of the homozygotes produces a phenotype that survives, but the heterozygote is viable.

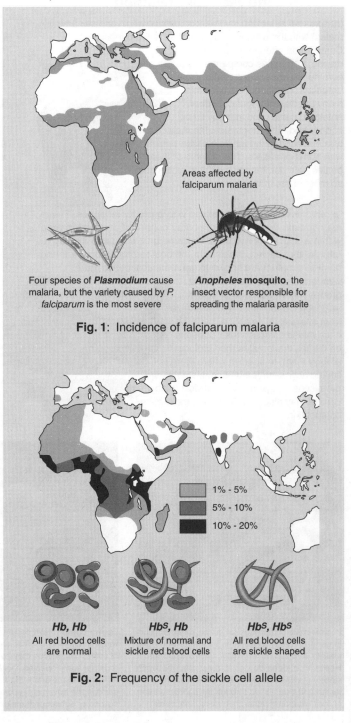

Four species of *Plasmodium* cause malaria, but the variety caused by *P. falciparum* is the most severe

Anopheles mosquito, the insect vector responsible for spreading the malaria parasite

Areas affected by falciparum malaria

Fig. 1: Incidence of falciparum malaria

1% - 5%
5% - 10%
10% - 20%

Hb, Hb
All red blood cells are normal

Hbˢ, Hb
Mixture of normal and sickle red blood cells

Hbˢ, Hbˢ
All red blood cells are sickle shaped

Fig. 2: Frequency of the sickle cell allele

1. With respect to the sickle cell allele, explain how **heterozygous advantage** can lead to **balanced polymorphism**:

_____

_____

_____

_____

_____

Mechanisms of Evolution

Code: EA 3

Sexual Selection

The success of an individual is measured not only by the number of offspring it leaves, but also by the quality or likely reproductive success of those offspring. This means that it becomes important who its mate will be. It was Darwin (1871) who first introduced the concept of sexual selection; a special type of natural selection that produces anatomical and behavioural traits that affect an individual's ability to acquire mates. Biologists today recognise two types: **intrasexual selection** (usually male-male competition) and **intersexual selection** or mate selection. One result of either type is the evolution of **sexual dimorphism**.

Intrasexual Selection

Intrasexual selection involves competition within one sex (usually males) with the winner gaining access to the opposite sex. Competition often takes place before mating, and males compete to establish dominance or secure a territory for breeding or mating. This occurs in many species of ungulates (**deer**, antelope, cattle) and in many birds. In deer and other ungulates, the males typically engage in highly ritualised battles with horns or antlers. The winners of these battles gain dominance over rival males and do most of the mating.

In other species, males compete vigorously for territories. These may contain resources or they may consist of an isolated area within a special arena used for communal courtship display (a **lek**). In lek species, males with the best territories on a lek (the dominant males) are known to get more chances to mate with females. In some species of grouse (right), this form of sexual selection can be difficult to distinguish from intersexual selection, because once males establish their positions on the lek the females then choose among them. In species where there is limited access to females and females are promiscuous, **sperm competition** (below, centre) may also be a feature of male-male competition.

Intersexual Selection

In intersexual selection (or **mate choice**), individuals of one sex (usually the males) advertise themselves as potential mates and members of the other sex (usually the females), choose among them. Intersexual selection results in development of exaggerated ornamentation, such as elaborate plumages. Female preference for elaborate male ornaments is well supported by both anecdotal and experimental evidence. For example, in the **long-tailed widow bird** (*Euplectes progne*), females prefer males with long tails. When tails are artificially shortened or lengthened, females still prefer males with the longest tails; they therefore select for long tails, not another trait correlated with long tails.

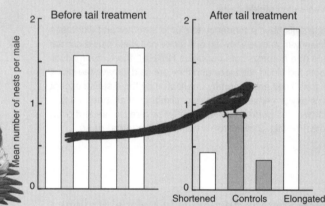

As shown above, there was no significant difference in breeding success between the groups before the tails were altered. When the tails were cut and lengthened, breeding success went down and up respectively.

In male-male competition for mates, ornamentation is used primarily to advertise superiority to rival males, and not to mortally wound opponents. However, injuries do occur, most often between closely matched rivals, where dominance must be tested and established through the aggressive use of their weaponry rather than mere ritual duels.

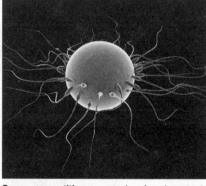

Sperm competition occurs when females remate within a relatively short space of time. The outcome of sperm competition may be determined by mating order. In some species, including those that guard their mates, the first male has the advantage, but in many the advantage accrues to the sperm of the second or subsequent males.

How do male features, such as the extravagant plumage of the peacock, persist when increasingly elaborate plumage must become detrimental to survival at some point? At first, preference for such traits must confer a survival advantage. Male adornment and female preference then advance together until a stable strategy is achieved.

1. Explain the difference between **intrasexual selection** and **mate selection**, identifying the features associated with each:

_____

_____

2. Suggest how sexual selection results in marked **sexual dimorphism**: _____

_____

_____

Darwin's Finches

The Galapagos Islands, 920 km off the west coast of Ecuador, played a major role in shaping Darwin's thoughts about evolution. While exploring the islands in 1835, he was struck by the unique and peculiar species he found there, in particular, the island's finches. The Galapagos group is home to 13 species of finches in four genera. This variety has arisen as a result of evolution from one ancestral species. Initially, a number of small finches, probably grassquits, made their way from South America to the Galapagos Islands. In the new environment, which was relatively free of competitors, the colonisers underwent an adaptive radiation, producing a range of species each with its own unique feeding niche. Although similar in their plumage, nest building techniques, and calls, the different species can be distinguished by the size and shape of their beaks. Each species has a beak adapted for a different purpose, such as crushing seeds, pecking wood, or probing flowers. Between them, the 13 species of this endemic group fill the roles of seven different families of South American mainland birds. DNA analyses have confirmed Darwin's insight and have shown that all 13 species evolved from a flock of about 30 birds arriving a million years ago.

The Evolution of Darwin's Finches

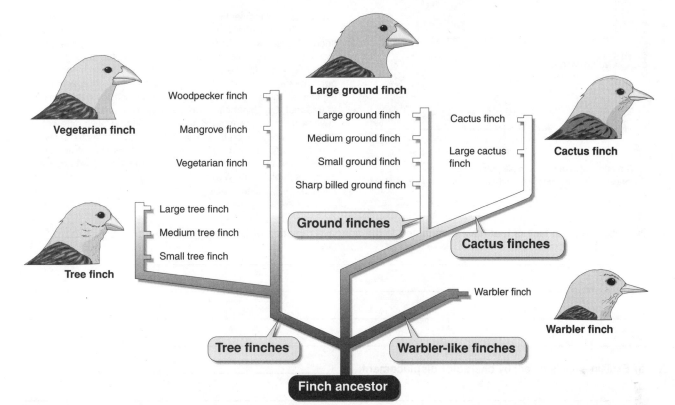

Vegetarian finch

Woodpecker finch

Mangrove finch

Vegetarian finch

Large ground finch

Large ground finch

Medium ground finch

Small ground finch

Sharp billed ground finch

Cactus finch

Large cactus finch

Cactus finch

Ground finches

Cactus finches

Large tree finch

Medium tree finch

Small tree finch

Tree finch

Warbler finch

Warbler finch

Tree finches

Warbler-like finches

Finch ancestor

Small tree finch

Large tree finch

As the name implies, **tree finches** are largely arboreal and feed mainly on insects. The bill is sharper than in ground finches and better suited to grasping insects. Paler than ground or cactus finches, they also have streaked breasts.

There are four species of **ground finches** with crushing-type bills used for seed eating. On Wolf Island, they are called vampire finches because they peck the skin of animals to draw blood, which they then drink (see left).

Cactus finches have most probably descended from ground finches. They have a probing beak and feed on insects on the cactus or the cactus itself. On islands where there are no ground finches, there is more variation in beak size than on the islands where the species coexist.

The **warbler finch** is named for its resemblance to the unrelated warblers. The beak of the warbler finch is the thinnest of the Galapagos finches. It is also the most widespread species, found throughout the archipelago. Warbler finches prey on flying and ground dwelling insects.

California Academy of Sciences

Mechanisms of Evolution

Code: A 2

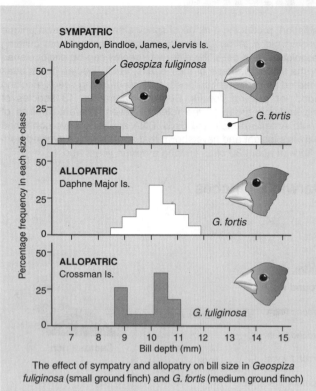

SYMPATRIC
Abingdon, Bindloe, James, Jervis Is.

Geospiza fuliginosa

G. fortis

ALLOPATRIC
Daphne Major Is.

G. fortis

ALLOPATRIC
Crossman Is.

G. fuliginosa

Percentage frequency in each size class

Bill depth (mm)

The effect of sympatry and allopatry on bill size in *Geospiza fuliginosa* (small ground finch) and *G. fortis* (medium ground finch)

Sympatry and Character Displacement

There is good evidence that finch evolution appears to be driven by a combination of allopatric and sympatric events. Coexisting species of ground finches on four islands (top graph) show large differences in bill sizes, enabling each species to feed on different sized seeds. However when either species exists in the absence of the other on different islands (lower graphs), it possesses intermediate bill sizes (about 10 mm) enabling it to feed without partitioning seed resources. This phenomenon, whereby competition causes two closely related species to become more different in regions where their ranges overlap, is referred to as **character displacement**.

Character displacement is evident in other populations of finches as well. There are well-studied populations of the large cactus finch (*G. conirostris*) on Genovesa and Espanola Islands, but their bill sizes are quite different. On Genovesa, the large ground finch coexists with the large cactus finch. In these sympatric populations, the variability in bill size *within* each species is minimal but there is little overlap *between* the species with respect to this trait. On Espanola, where the large ground finch either never arrived, or became extinct, the situation is quite different. With no competition on Espanola, the large cactus finch displays a greater variability in bill size. Its bill is somewhat intermediate between the two finches on Genovesa, and it can feed equally well in both niches all year round.

Data based on an adaptation by Strickberger (2000)

1. Describe the main factors that have contributed to the adaptive radiation of Darwin's finches: _____

_____

_____

_____

_____

2. (a) Explain what is meant by **character displacement**: _____

_____

_____

(b) Discuss how the incidence of character displacement observed in the Galapagos finches supports the view that their adaptive radiation from a common ancestor has been driven by a combination of allopatric and sympatric events:

_____

_____

_____

_____

3. The range of variability shown by a phenotype in response to environmental variation is called **phenotypic plasticity**.

(a) Discuss the evidence for phenotypic plasticity in Galapagos finches: _____

_____

_____

_____

(b) Explain what this suggests about the biology of the original finch ancestor: _____

_____

_____

Gene Pools and Evolution

The diagram below illustrates the dynamic nature of **gene pools**. It portrays two imaginary populations of one beetle species. Each beetle is a 'carrier' of genetic information, represented here by the alleles (A and a) for a single **codominant gene** that controls the beetle's colour. Normally, there are three versions of the phenotype: black, dark, and pale. Mutations may create other versions of the phenotype. Some of the **microevolutionary processes** that can affect the genetic composition (**allele frequencies**) of the gene pool are illustrated. See the activity *Gene Pool Exercise* for cut-out beetles to simulate this activity.

Immigration: Populations can gain alleles when they are introduced from other gene pools. Immigration is one aspect of gene flow.

Mutations: Spontaneous mutations can develop that alter the allele frequencies of the gene pool, and even create new alleles. Mutation is very important to evolution, because it is the original source of genetic variation that provides new material for natural selection.

Emigration: Genes may be lost to other gene pools.

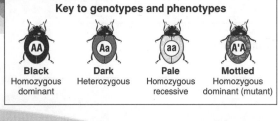

Deme 1

The term deme describes a local population that is genetically isolated from other populations in the species. Demes usually have some clearly definable genetic or other character that sets them apart from other populations.

Natural selection: Selection pressure against certain allele combinations may reduce reproductive success or lead to death. Natural selection sorts genetic variability, and accumulates and maintains favourable genotypes in a population. It tends to reduce genetic diversity within the gene pool and increase differences between populations.

Geographical barriers: Isolate the gene pool and prevent *regular* gene flow between populations.

Gene flow: Genes are exchanged with other gene pools as individuals move between them. Gene flow is a source of new genetic variation and tends to reduce differences between populations that have accumulated because of natural selection or genetic drift.

Key to genotypes and phenotypes

Black	**Dark**	**Pale**	**Mottled**
Homozygous dominant	Heterozygous	Homozygous recessive	Homozygous dominant (mutant)

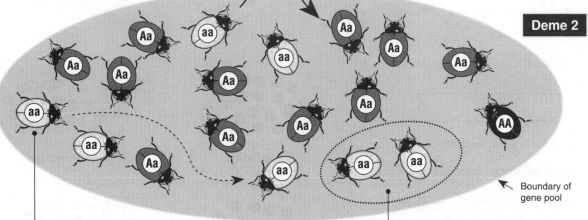

Deme 2

Boundary of gene pool

Mate selection (non-random mating): Individuals may not select their mate randomly and may seek out particular phenotypes, increasing the frequency of these "favoured" alleles in the population.

Genetic drift: Chance events can cause the allele frequencies of small populations to "drift" (change) randomly from generation to generation. Genetic drift can play a significant role in the microevolution of very small populations. The two situations most often leading to populations small enough for genetic drift to be significant are the *bottleneck effect* (where the population size is dramatically reduced by a catastrophic event) and the *founder effect* (where a small number of individuals colonise a new area).

Mechanisms of Evolution

Code: PA 2

1. For each of the 2 demes shown on the previous page (treating the mutant in deme 1 as a AA):

   (a) Count up the numbers of **allele types** (**A** and **a**).

   (b) Count up the numbers of **allele combinations** (**AA, Aa, aa**).

2. Calculate the frequencies as percentages (%) for the allele types and combinations:

Deme 1		Number counted	%
Allele types	A		
	a		
Allele combinations	AA		
	Aa		
	aa		

Deme 2		Number counted	%
Allele types	A		
	a		
Allele combinations	AA		
	Aa		
	aa		

3. One of the fundamental concepts for population genetics is that of **genetic equilibrium**, stated as: *"For a very large, randomly mating population, the proportion of dominant to recessive alleles remains constant from one generation to the next"*. If a gene pool is to remain unchanged, it must satisfy all of the criteria below that favour gene pool stability. Few populations meet all (or any) of these criteria and their genetic makeup must therefore by continually changing. For each of the five factors (a-e) below, state briefly **how** and **why** each would affect the allele frequency in a gene pool:

   (a) Population size: _____

   _____

   _____

   _____

   (b) Mate selection: _____

   _____

   _____

   _____

   (c) Gene flow between populations: _____

   _____

   _____

   _____

   (d) Mutations: _____

   _____

   _____

   _____

   (e) Natural selection: _____

   _____

   _____

   _____

4. Identify the factors that tend to:

   (a) Increase genetic variation in populations:

   _____

   _____

   (b) Decrease genetic variation in populations:

   _____

   _____

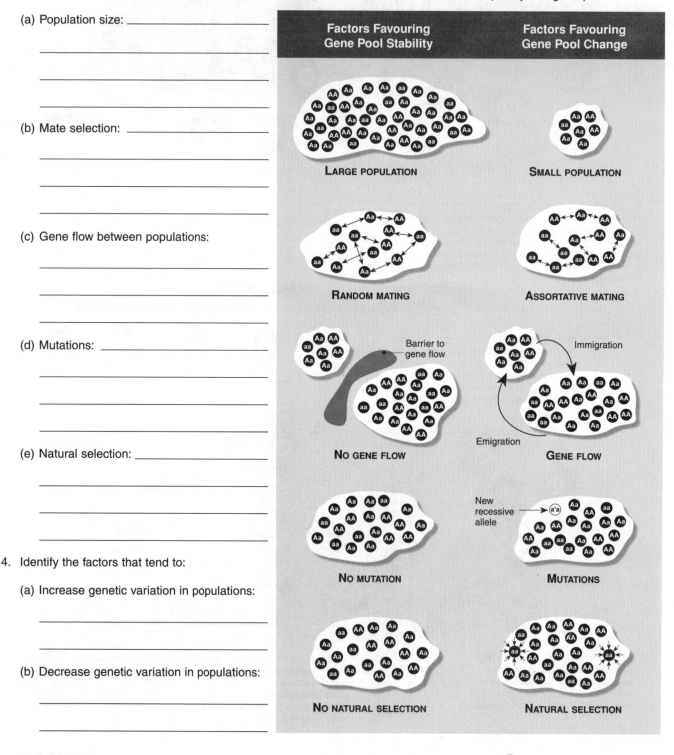

Factors Favouring Gene Pool Stability / Factors Favouring Gene Pool Change

LARGE POPULATION — SMALL POPULATION

RANDOM MATING — ASSORTATIVE MATING

NO GENE FLOW (Barrier to gene flow) — GENE FLOW (Immigration / Emigration)

NO MUTATION — MUTATIONS (New recessive allele)

NO NATURAL SELECTION — NATURAL SELECTION

Gene Pool Exercise

Cut out each of the beetles on this page and use them to simulate different events within a gene pool as described in this topic (pages: *Gene Pools and Evolution, Changes in a Gene Pool, The Founder Effect, Population Bottlenecks, Genetic Drift*).

Mechanisms of Evolution

Code: P 3

This page has deliberately been left blank

Population Genetics Calculations

The **Hardy-Weinberg equation** provides a simple mathematical model of genetic equilibrium in a gene pool, but its main application in population genetics is in calculating allele and genotype frequencies in populations, particularly as a means of studying changes and measuring their rate. The use of the Hardy-Weinberg equation is described below.

Punnett square

A   a

	A	a
A	AA	Aa
a	aA	aa

Frequency of allele combination **AA** in the population is represented as p^2

Frequency of allele combination **aa** in the population is represented as q^2

Frequency of allele combination **Aa** in the population (add these together to get **2pq**)

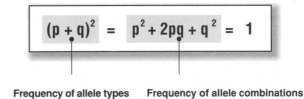

$$(p + q)^2 \;=\; p^2 + 2pq + q^2 \;=\; 1$$

Frequency of allele types

p = Frequency of allele A

q = Frequency of allele a

Frequency of allele combinations

p^2 = Frequency of AA (homozygous dominant)

2pq = Frequency of Aa (heterozygous)

q^2 = Frequency of aa (homozygous recessive)

The Hardy-Weinberg equation is applied to populations with a simple genetic situation: dominant and recessive alleles controlling a single trait. The frequency of all of the dominant (A) and recessive alleles (a) equals the total genetic complement, and adds up to 1 or 100% of the alleles present.

How To Solve Hardy-Weinberg Problems

In most populations, the frequency of two alleles of interest is calculated from the proportion of homozygous recessives (q^2), as this is the only genotype identifiable directly from its phenotype. If only the dominant phenotype is known, q^2 may be calculated (1 – the frequency of the dominant phenotype). The following steps outline the procedure for solving a Hardy-Weinberg problem:

Remember that all calculations must be carried out using proportions, NOT PERCENTAGES!

1. Examine the question to determine what piece of information you have been given about the population. In most cases, this is the percentage or frequency of the homozygous recessive phenotype q^2, or the dominant phenotype $p^2 + 2pq$ (see note above).

2. The first objective is to find out the value of p or q, If this is achieved, then every other value in the equation can be determined by simple calculation.

3. Take the square root of q^2 to find q.

4. Determine p by subtracting q from 1 (i.e. p = 1 – q).

5. Determine p^2 by multiplying p by itself (i.e. $p^2 = p \times p$).

6. Determine 2pq by multiplying p times q times 2.

7. Check that your calculations are correct by adding up the values for $p^2 + q^2 + 2pq$ (the sum should equal 1 or 100%).

Worked example

In the American white population approximately 70% of people can taste the chemical phenylthiocarbamide (PTC) (the dominant phenotype), while 30% are non-tasters (the recessive phenotype).

Determine the frequency of: **Answers**

(a) Homozygous recessive phenotype(q^2).   30% - provided

(b) The dominant allele (**p**).   45.2%

(c) Homozygous tasters (**p^2**).   20.5%

(d) Heterozygous tasters (**2pq**).   49.5%

Data: The frequency of the dominant phenotype (70% tasters) and recessive phenotype (30% non-tasters) are provided.

Working:

Recessive phenotype: q^2 = 30%
use 0.30 for calculation

therefore: q = 0.5477
square root of 0.30

therefore: p = 0.4523
1 – q = p
1 – 0.5477 = 0.4523

Use p and q in the equation (top) to solve any unknown:

Homozygous dominant p^2 = 0.2046
(p x p = 0.4523 x 0.4523)

Heterozygous: 2pq = 0.4953

---

1. A population of hamsters has a gene consisting of 90% M alleles (black) and 10% m alleles (grey). Mating is random.

   Data: Frequency of recessive allele (10% m) and dominant allele (90% M).

   Determine the proportion of offspring that will be black and the proportion that will be grey (show your working).

   _____

   _____

Recessive allele:	q =	
Dominant allele:	p =	
Recessive phenotype:	q^2 =	
Homozygous dominant:	p^2 =	
Heterozygous:	2pq =	

Mechanisms of Evolution

Code: RDA 3

2. You are working with pea plants and found 36 plants out of 400 were dwarf.
   Data: Frequency of recessive phenotype (36 out of 400 = 9%)

   (a) Calculate the frequency of the tall gene: _____

   (b) Determine the number of heterozygous pea plants:

   _____

   _____

Recessive allele:	q =	
Dominant allele:	p =	
Recessive phenotype:	q^2 =	
Homozygous dominant:	p^2 =	
Heterozygous:	2pq =	

3. In humans, the ability to taste the chemical phenylthiocarbaminde (PTC) is inherited as a simple dominant characteristic. Suppose you found out that 360 out of 1000 college students could not taste the chemical.
   Data: Frequency of recessive phenotype (360 out of 1000).

   (a) State the frequency of the gene for tasting PTC:

   _____

   (b) Determine the number of heterozygous students in this population:

   _____

Recessive allele:	q =	
Dominant allele:	p =	
Recessive phenotype:	q^2 =	
Homozygous dominant:	p^2 =	
Heterozygous:	2pq =	

4. A type of deformity appears in 4% of a large herd of cattle. Assume the deformity was caused by a recessive gene.
   Data: Frequency of recessive phenotype (4% deformity).

   (a) Calculate the percentage of the herd that are carriers of the gene:

   _____

   (b) Determine the frequency of the dominant gene in this case:

   _____

Recessive allele:	q =	
Dominant allele:	p =	
Recessive phenotype:	q^2 =	
Homozygous dominant:	p^2 =	
Heterozygous:	2pq =	

5. Assume you placed 50 pure bred black guinea pigs (dominant allele) with 50 albino guinea pigs (recessive allele) and allowed the population to attain genetic equilibrium (several generations have passed).
   Data: Frequency of recessive allele (50%) and dominant allele (50%).

   Determine the proportion (%) of the population that becomes white:

   _____

Recessive allele:	q =	
Dominant allele:	p =	
Recessive phenotype:	q^2 =	
Homozygous dominant:	p^2 =	
Heterozygous:	2pq =	

6. It is known that 64% of a large population exhibit the recessive trait of a characteristic controlled by two alleles (one is dominant over the other).
   Data: Frequency of recessive phenotype (64%). Determine the following:

   (a) The frequency of the recessive allele: _____

   (b) The percentage that are heterozygous for this trait: _____

   (c) The percentage that exhibit the dominant trait: _____

   (d) The percentage that are homozygous for the dominant trait: _____

   (e) The percentage that has one or more recessive alleles: _____

7. Albinism is recessive to normal pigmentation in humans. The frequency of the albino allele was 10% in a population.
   Data: Frequency of recessive allele (10% albino allele).

   Determine the proportion of people that you would expect to be albino:

   _____

   _____

Recessive allele:	q =	
Dominant allele:	p =	
Recessive phenotype:	q^2 =	
Homozygous dominant:	p^2 =	
Heterozygous:	2pq =	

Analysis of a Squirrel Gene Pool

In Olney, Illinois, in the United States, there is a unique population of albino (white) and grey squirrels. Between 1977 and 1990, students at Olney Central College carried out a study of this population. They recorded the frequency of grey and albino squirrels. The albinos displayed a mutant allele expressed as an albino phenotype only in the homozygous recessive condition. The data they collected are provided in the table below. Using the **Hardy-Weinberg equation** for calculating genotype frequencies, it was possible to estimate the frequency of the normal 'wild' allele (G) providing grey fur colouring, and the frequency of the mutant albino allele (g) producing white squirrels. This study provided real, first hand, data that students could use to see how genotype frequencies can change in a real population.

Thanks to **Dr. John Stencel**, Olney Central College, Olney, Illinois, US, for providing the data for this exercise.

Grey squirrel, usual colour form   Albino form of grey squirrel

Population of grey and white squirrels in Olney, Illinois (1977-1990)

Year	Grey	White	Total	GG	Gg	gg	Freq. of g	Freq. of G
1977	602	182	784	26.85	49.93	23.21	48.18	51.82
1978	511	172	683	24.82	50.00	25.18	50.18	49.82
1979	482	134	616	28.47	49.77	21.75	46.64	53.36
1980	489	133	622	28.90	49.72	21.38	46.24	53.76
1981	536	163	699	26.74	49.94	23.32	48.29	51.71
1982	618	151	769	31.01	49.35	19.64	44.31	55.69
1983	419	141	560	24.82	50.00	25.18	50.18	49.82
1984	378	106	484	28.30	49.79	21.90	46.80	53.20
1985	448	125	573	28.40	49.78	21.82	46.71	53.29
1986	536	155	691	27.71	49.86	22.43	47.36	52.64
1987	No data collected this year							
1988	652	122	774	36.36	47.88	15.76	39.70	60.30
1989	552	146	698	29.45	49.64	20.92	45.74	54.26
1990	603	111	714	36.69	47.76	15.55	39.43	60.57

1. **Graph population changes**: Use the data in the first three columns of the table above to plot a line graph. This will show changes in the phenotypes: numbers of grey and white (albino) squirrels, as well as changes in the total population. Plot: **grey**, **white**, and **total** for each year:

    (a) By how much have total population numbers fluctuated over the sampling period (as a %):

    _____

    (b) Describe the overall trend in total population numbers and any pattern that may exist:

    _____

    _____

    _____

    _____

    _____

    _____

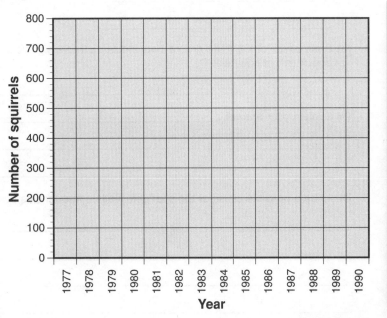

Mechanisms of Evolution

Code: EDA 3

2. **Graph genotype changes**: Use the data in the genotype columns of the table on the previous page to plot a line graph. This will show changes in the allele combinations (**GG**, **Gg**, **gg**). Plot: **GG**, **Gg**, and **gg** for each year:

Describe the overall trend in the frequency of:

(a) Homozygous dominant (**GG**) genotype:

_____

_____

(b) Heterozygous (**Gg**) genotype:

_____

_____

(c) Homozygous recessive (**gg**) genotype:

_____

_____

_____

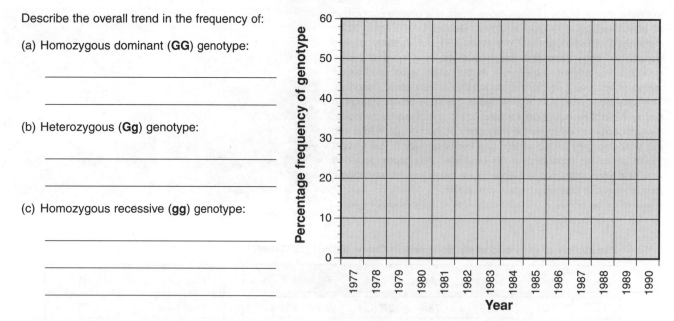

3. **Graph allele changes**: Use the data in the last two columns of the table on the previous page to plot a line graph. This will show changes in the *allele frequencies* for each of the dominant (**G**) and recessive (**g**) alleles.
Plot: the frequency of **G** and the frequency of **g**:

(a) Describe the overall trend in the frequency of the dominant allele (**G**):

_____

_____

_____

(b) Describe the overall trend in the frequency of the recessive allele (**g**):

_____

_____

_____

_____

_____

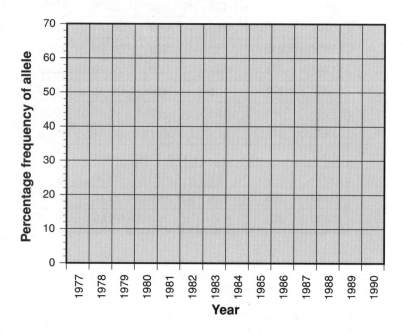

4. (a) State which of the three graphs best indicates that a significant change may be taking place in the gene pool of this population of squirrels:

_____

(b) Explain your answer: _____

_____

_____

5. Describe a possible cause of the changes in allele frequencies over the sampling period: _____

_____

_____

_____

Changes in a Gene Pool

The diagram below shows an imaginary population of beetles undergoing changes as it is subjected to two 'events'. The three phases represent a progression in time, i.e. the same gene pool, undergoing change. The beetles have three phenotypes determined by the amount of pigment deposited in the cuticle. Three versions of this trait exist: black, dark, and pale. The gene controlling this character is represented by two alleles **A** and **a**. Your task is to analyse the gene pool as it undergoes changes.

Phase 1: Initial gene pool

Calculate the frequencies of the *allele types* and *allele combinations* by counting the actual numbers, then working them out as percentages.

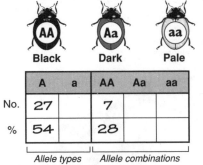

Black   Dark   Pale

	A	a	AA	Aa	aa
No.	27		7		
%	54		28		

Allele types   Allele combinations

Phase 2: Natural selection

In the same gene pool at a later time there was a change in the allele frequencies. This was due to the loss of certain allele combinations due to natural selection. Some of those with a genotype of aa were eliminated (poor fitness).

Calculate as for above. Do not include the individuals surrounded by small white arrows in your calculations; they are dead!

	A	a	AA	Aa	aa
No.					
%					

Phase 3: Immigration and emigration

This particular kind of beetle exhibits wandering behaviour. The allele frequencies change again due to the introduction and departure of individual beetles, each carrying certain allele combinations.

Calculate as above. In your calculations, include the individual coming into the gene pool (AA), but remove the one leaving (aa).

	A	a	AA	Aa	aa
No.					
%					

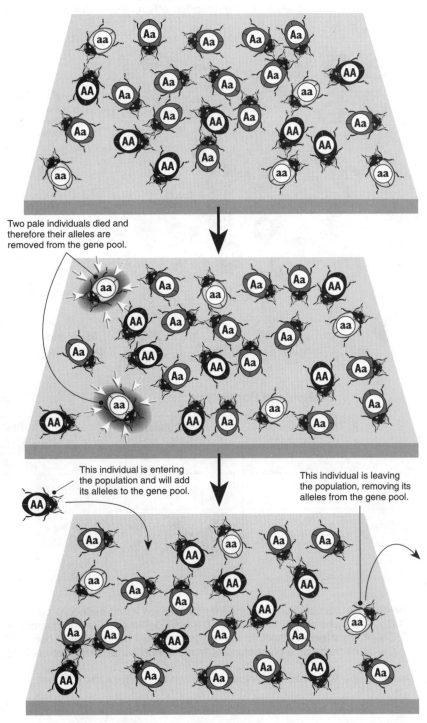

Two pale individuals died and therefore their alleles are removed from the gene pool.

This individual is entering the population and will add its alleles to the gene pool.

This individual is leaving the population, removing its alleles from the gene pool.

1. Explain how the number of dominant alleles (A) in the genotype of a beetle affects its phenotype:

_____

_____

2. For each phase in the gene pool above (place your answers in the tables provided):
   (a) Determine the relative frequencies of the two alleles: A and a. Simply total the **A** alleles and **a** alleles separately.
   (b) Determine the frequency of how the alleles come together as allele pair combinations in the gene pool (AA, Aa and aa). Count the number of each type of combination.
   (c) For each of the above, work out the frequencies as percentages:

   Allele frequency = No. counted alleles / total no. of alleles x 100

Code: PDA 3

Mechanisms of Evolution

The Founder Effect

Occasionally, a small number of individuals from a large population may migrate away, or become isolated from, their original population. If this colonising or 'founder' population is made up of only a few individuals, it will probably have a *non-representative sample* of alleles from the parent population's gene pool. As a consequence of this **founder effect**, the colonising population may evolve differently from that of the parent population, particularly since the environmental conditions for the isolated population may be different. In some cases, it may be possible for certain alleles to be missing altogether from the individuals in the isolated population. Future generations of this population will not have this allele.

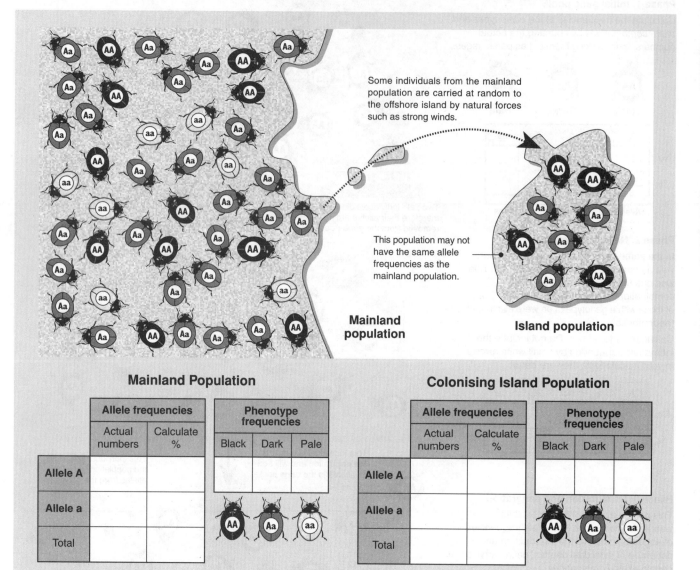

Mainland Population

	Allele frequencies		Phenotype frequencies		
	Actual numbers	Calculate %	Black	Dark	Pale
Allele A					
Allele a					
Total					

Colonising Island Population

	Allele frequencies		Phenotype frequencies		
	Actual numbers	Calculate %	Black	Dark	Pale
Allele A					
Allele a					
Total					

1. Compare the mainland population to the population which ended up on the island (use the spaces in the tables above):
   (a) Count the **phenotype** numbers for the two populations (i.e. the number of black, dark and pale beetles).
   (b) Count the **allele** numbers for the two populations: the number of dominant alleles (A) and recessive alleles (a). Calculate these as a percentage of the total number of alleles for each population.

2. Describe how the allele frequencies of the two populations are different: _____

   _____

3. Describe some possible ways in which various types of organism can be carried to an offshore island:

   (a) Plants: _____

   (b) Land animals: _____

   (c) Non-marine birds: _____

4. Since founder populations are often very small, describe another process that may further alter the allele frequencies:

   _____

   _____

Population Bottlenecks

Populations may sometimes be reduced to low numbers by predation, disease, or periods of climatic change. A population crash may not be 'selective': it may affect all phenotypes equally. Large scale catastrophic events, such as fire or volcanic eruption, are examples of such non-selective events. Humans may severely (and selectively) reduce the numbers of some species through hunting and/or habitat destruction. These populations may recover, having squeezed through a 'bottleneck' of low numbers. The

diagram below illustrates how population numbers may be reduced as a result of a catastrophic event. Following such an event, the small number of individuals contributing to the gene pool may not have a representative sample of the genes in the pre-catastrophe population, i.e. the allele frequencies in the remnant population may be severely altered. Genetic drift may cause further changes to allele frequencies. The small population may return to previous levels but with a reduced genetic diversity.

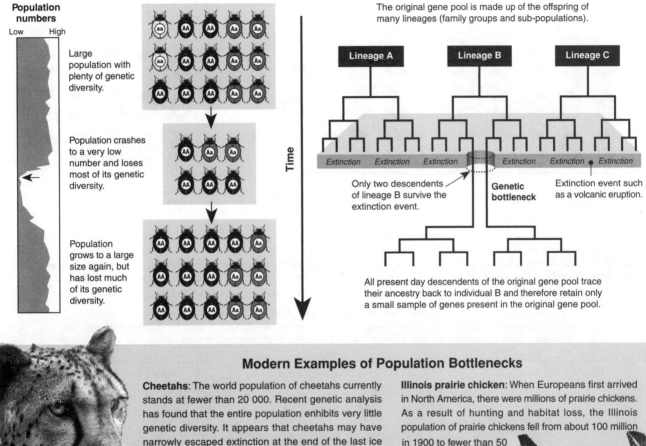

Population numbers

Low   High

Large population with plenty of genetic diversity.

Population crashes to a very low number and loses most of its genetic diversity.

Population grows to a large size again, but has lost much of its genetic diversity.

Time

The original gene pool is made up of the offspring of many lineages (family groups and sub-populations).

Lineage A   Lineage B   Lineage C

Extinction   Extinction   Extinction   Extinction   Extinction   Extinction

Only two descendents of lineage B survive the extinction event.

Genetic bottleneck

Extinction event such as a volcanic eruption.

All present day descendents of the original gene pool trace their ancestry back to individual B and therefore retain only a small sample of genes present in the original gene pool.

Modern Examples of Population Bottlenecks

Cheetahs: The world population of cheetahs currently stands at fewer than 20 000. Recent genetic analysis has found that the entire population enhibits very little genetic diversity. It appears that cheetahs may have narrowly escaped extinction at the end of the last ice age, about 10-20 000 years ago. If all modern cheetahs arose from a very limited genetic stock, this would explain their present lack of genetic diversity. The lack of genetic variation has resulted in a number of problems that threaten cheetah survival, including sperm abnormalities, decreased fecundity, high cub mortality, and sensitivity to disease.

Illinois prairie chicken: When Europeans first arrived in North America, there were millions of prairie chickens. As a result of hunting and habitat loss, the Illinois population of prairie chickens fell from about 100 million in 1900 to fewer than 50 in the 1990s. A comparison of the DNA from birds collected in the mid-twentieth century and DNA from the surviving population indicated that most of the genetic diversity has been lost.

Photo: Dept. of Natural Resources, Illinois

1. Endangered species are often subjected to population bottlenecks. Explain how population bottlenecks affect the ability of a population of an endangered species to recover from its plight:

_____

_____

_____

2. Explain why the lack of genetic diversity in cheetahs has increased their sensitivity to disease: _____

_____

_____

3. Describe the effect of a population bottleneck on the potential of a species to adapt to changes (i.e. its ability to evolve):

_____

_____

Mechanisms of Evolution

Code: PA 3

Genetic Drift

Not all individuals, for various reasons, will be able to contribute their genes to the next generation. **Genetic drift** (also known as the Sewall-Wright Effect) refers to the *random changes in allele frequency* that occur in all populations, but are much more pronounced in small populations. In a small population, the effect of a few individuals not contributing their alleles to the next generation can have a great effect on allele frequencies. Alleles may even become **lost** from the gene pool altogether (frequency becomes 0%) or **fixed** as the only allele for the gene present (frequency becomes 100%).

The genetic makeup (allele frequencies) of the population changes randomly over a period of time

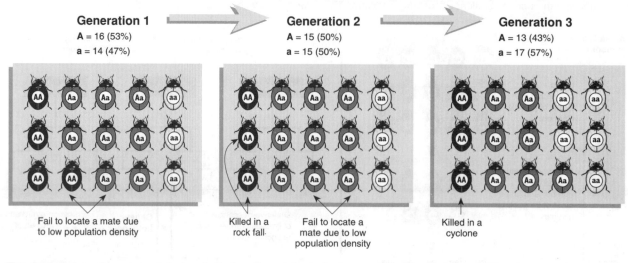

Generation 1
A = 16 (53%)
a = 14 (47%)

Generation 2
A = 15 (50%)
a = 15 (50%)

Generation 3
A = 13 (43%)
a = 17 (57%)

Fail to locate a mate due to low population density

Killed in a rock fall

Fail to locate a mate due to low population density

Killed in a cyclone

This diagram shows the gene pool of a hypothetical small population over three generations. For various reasons, not all individuals contribute alleles to the next generation. With the random loss of the alleles carried by these individuals, the allele frequency changes from one generation to the next. The change in frequency is directionless as there is no selecting force. The allele combinations for each successive generation are determined by how many alleles of each type are passed on from the preceding one.

Computer Simulation of Genetic Drift

Below are displayed the change in allele frequencies in a computer simulation showing random genetic drift. The breeding population progressively gets smaller from left to right. Each simulation was run for 140 generations.

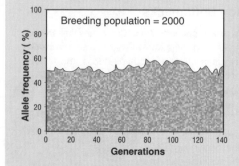

Breeding population = 2000

Breeding population = 200

Breeding population = 20

Allele lost from the gene pool

Large breeding population
Fluctuations are minimal in large breeding populations because the large numbers buffer the population against random loss of alleles. On average, losses for each allele type will be similar in frequency and little change occurs.

Small breeding population
Fluctuations are more severe in smaller breeding populations because random changes in a few alleles cause a greater percentage change in allele frequencies.

Very small breeding population
Fluctuations in very small breeding populations are so extreme that the allele can become fixed (frequency of 100%) or lost from the gene pool altogether (frequency of 0%).

1. Explain what is meant by **genetic drift**: _____

   _____

2. Explain how genetic drift affects the amount of genetic variation within very small populations: _____

   _____

   _____

   _____

3. Identify a small breeding population of animals or plants in your country in which genetic drift could be occurring:

   _____

Artificial Selection

The ability of people to control the breeding of domesticated animals and crop plants has resulted in an astounding range of phenotypic variation over relatively short time periods. Most agricultural plants and animals, as well as pets, have undergone **artificial selection** (selective breeding). The dog is a striking example of this, as there are now over 400 different breeds. Artificial selection involves breeding from individuals with the most desirable phenotypes. The aim of this is to alter the average phenotype within the species. As well as selecting for physical characteristics, desirable behavioural characteristics (e.g. the ability to 'read' the body language of humans) has also been selected for in dogs. All breeds of dog are members of the same species, *Canis familiaris*. This species descended from a single wild species, the grey wolf *Canis lupus*, over 15 000 years ago. Five ancient dog breeds are recognised, from which all other breeds are thought to have descended by artificial selection.

The Ancestor of Domestic Dogs

Until recently, it was unclear whether the ancestor to the modern domestic dogs was the desert wolf of the Middle East, the woolly wolf of central Asia, or the grey wolf of Northern Hemisphere. Recent genetic studies (mitochondrial DNA comparisons) now provide strong evidence that the ancestor of domestic dogs throughout the world is the grey wolf. It seems likely that this evolutionary change took place in a single region, most probably China.

Grey wolf *Canis lupus pallipes*

The grey wolf is distributed throughout Europe, North America, and Asia. Amongst members of this species, there is a lot of variation in coat coloration. This accounts for the large variation in coat colours of dogs today.

Dogs introduced to North America by humans 10 000 to 15 000 years ago

The first dog breeds probably originated in China at least 15 000 years ago, later spreading to other parts of the world

Mastiff-type
Canis familiaris inostranzevi
Originally from Tibet, the first records of this breed of dog go back to the Stoneage.

Greyhound
Canis familiaris leineri
Drawings of this breed on pottery dated from 8000 years ago in the Middle East make it one of the oldest.

Pointer-type
Canis familiaris intermedius
Probably derived from the greyhound breed for the purpose of hunting small game.

Sheepdog
Canis familiaris metris optimae
Originating in Europe, this breed has been used to guard flocks from predators for thousands of years.

Wolf-like
Canis familiaris palustris
Found in snow covered habitats in northern Europe, Asia (Siberia), and North America (Alaska).

1. Explain how artificial selection can result in changes in a gene pool over time: _____

_____

_____

_____

2. Describe the behavioural tendency of wolves that predisposed them to becoming a domesticated animal: _____

_____

3. List the physical and behavioural traits that would be desirable (selected for) in the following uses of a dog:

   (a) Hunting large game (e.g. boar and deer): _____

   (b) Game fowl dog: _____

   (c) Stock control (sheep/cattle dog): _____

   (d) Family pet (house dog): _____

   (e) Guard dog: _____

Mechanisms of Evolution

Code: RA 2

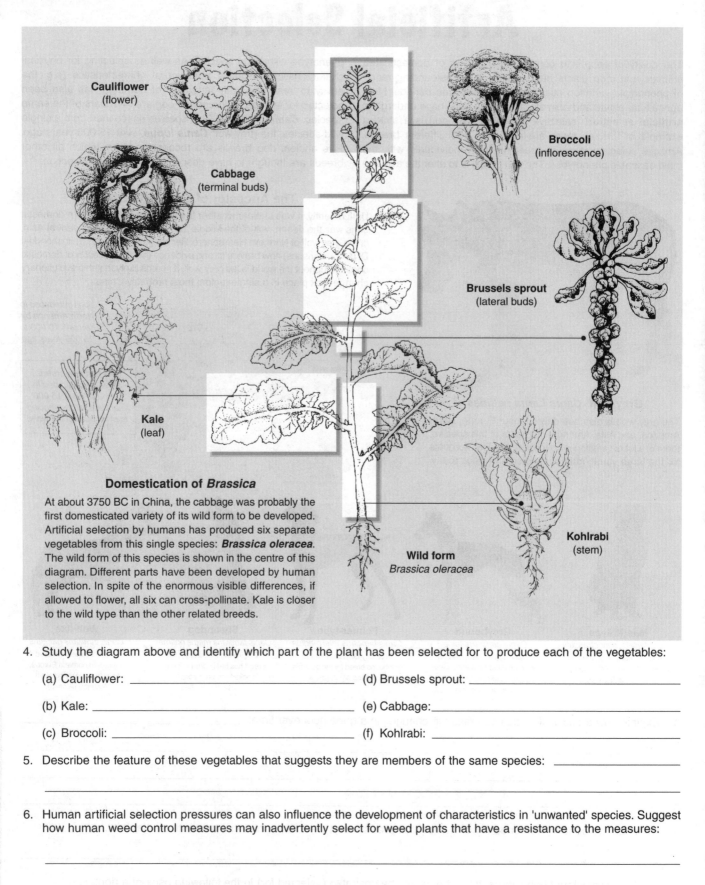

Cauliflower
(flower)

Cabbage
(terminal buds)

Broccoli
(inflorescence)

Brussels sprout
(lateral buds)

Kale
(leaf)

Kohlrabi
(stem)

Wild form
Brassica oleracea

Domestication of *Brassica*

At about 3750 BC in China, the cabbage was probably the first domesticated variety of its wild form to be developed. Artificial selection by humans has produced six separate vegetables from this single species: ***Brassica oleracea***. The wild form of this species is shown in the centre of this diagram. Different parts have been developed by human selection. In spite of the enormous visible differences, if allowed to flower, all six can cross-pollinate. Kale is closer to the wild type than the other related breeds.

4. Study the diagram above and identify which part of the plant has been selected for to produce each of the vegetables:

(a) Cauliflower: _____

(d) Brussels sprout: _____

(b) Kale: _____

(e) Cabbage: _____

(c) Broccoli: _____

(f) Kohlrabi: _____

5. Describe the feature of these vegetables that suggests they are members of the same species: _____

_____

6. Human artificial selection pressures can also influence the development of characteristics in 'unwanted' species. Suggest how human weed control measures may inadvertently select for weed plants that have a resistance to the measures:

_____

_____

_____

7. Explain how a farmer thousands of years ago was able to improve the phenotypic character of a cereal crop:

_____

_____

_____

Polyploidy in the Evolution of Wheat

Wheat has been cultivated for more than 9000 years and has undergone many changes during the process of its domestication. The evolution of wheat involved two natural hybridisation events, accompanied by **polyploidy**. **Hybrids** are the offspring of genetically dissimilar parents. They are important because they recombine the genetic characteristics of (often inbred) parental lines and show increased **heterozygosity**. This is associated with greater adaptability, survival, growth, and fertility in the offspring; a phenomenon known as **hybrid vigour** or heterosis. There is evidence to show that **interspecific hybridisation** (i.e. between species) was an important evolutionary mechanism in the domestication of wheat. **Polyploidy** has also played a major role in the evolution of crop plants. Most higher organisms are

diploid, i.e. two sets of chromosomes (2N), one set derived from each parent. If there are more than two sets, the organism is said to be **polyploid**. Diploids formed from hybridisation of genetically very dissimilar parents, e.g. from different species, are often infertile because the two sets of chromosomes are not able to pair properly at meiosis. In such hybrids, there are no gametes produced or the gametes are abnormal. In some cases of **allopolyploidy**, the chromosomes can be doubled and a tetraploid is formed from the diploid. This restores fertility to a hybrid, because each of the original chromosome sets can pair properly with each other during meiosis. These processes are outlined in the diagram below showing the history of domestication in wheat.

Polyploidy Events in the Evolution of Wheat

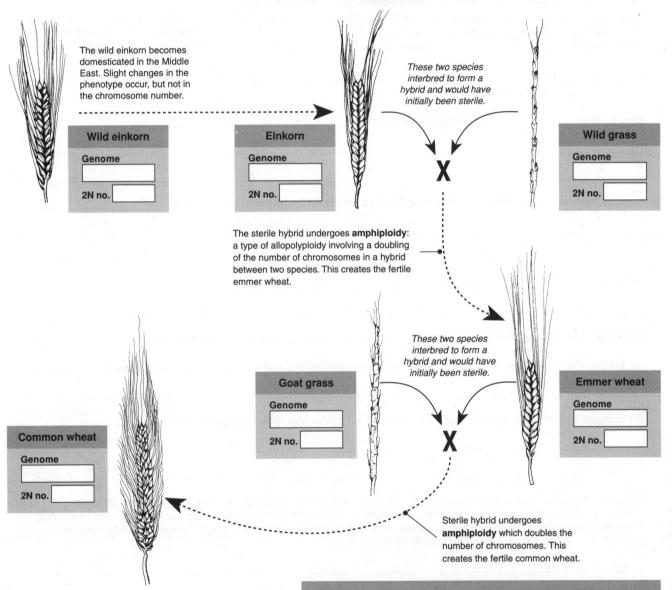

The wild einkorn becomes domesticated in the Middle East. Slight changes in the phenotype occur, but not in the chromosome number.

Wild einkorn

Genome

2N no.

Einkorn

Genome

2N no.

These two species interbred to form a hybrid and would have initially been sterile.

X

Wild grass

Genome

2N no.

The sterile hybrid undergoes **amphiploidy**: a type of allopolyploidy involving a doubling of the number of chromosomes in a hybrid between two species. This creates the fertile emmer wheat.

These two species interbred to form a hybrid and would have initially been sterile.

Goat grass

Genome

2N no.

X

Emmer wheat

Genome

2N no.

Common wheat

Genome

2N no.

Sterile hybrid undergoes **amphiploidy** which doubles the number of chromosomes. This creates the fertile common wheat.

The table on the right and the diagram above show the evolution of the common wheat. Common wheat is thought to have resulted from two sets of crossings between different species to produce hybrids. Wild einkorn (14 chromosomes, genome AA) evolved into einkorn, which crossed with a wild grass (14 chromosomes, genome BB) and gave rise to emmer wheat (28 chromosomes, genome AABB). Common wheat arose when emmer wheat was crossed with another type of grass (goat grass).

Common name	Species	Genome	Chromosomes N
Wild einkorn	*Triticum aegilopiodes*	**AA**	7
Einkorn	*Triticum monococcum*	**AA**	7
Wild grass	*Aegilops speltoides*	**BB**	7
Emmer wheat	*Triticum dicoccum*	**AABB**	14
Goat grass	*Aegilops squarrosa*	**DD**	7
Common wheat	*Triticum aestivum*	**AABBDD**	21

Mechanisms of Evolution

Code: A 3

Ancient cereal grasses had heads which shattered readily so that seeds would be scattered widely.

Modern wheat has been selected for its non shattering heads, high yield, and high gluten content.

Teosinte

Modern corn

Corn has also evolved during its domestication. Teosinte is thought to be the ancestor to both corn and maize.

1. Using the table on the previous page, label each of the wheats and grasses in the diagram with the correct **genome** and **2N** chromosome number for each plant.

2. Explain what is meant by **F₁ hybrid vigour** (heterosis): _____

_____

_____

3. Discuss the role of **polyploidy** and **interspecific hybridisation** in the evolution of wheat: _____

_____

_____

_____

_____

4. Cultivated wheat arose from wild, weedy ancestors through the selection of certain characters.

   (a) Identify the phenotypic traits that are desirable in modern wheat cultivars: _____

   _____

   (b) Suggest how ancient farmers would have carried out a selective breeding programme: _____

   _____

   _____

5. Cultivated American cotton plants have a total of 52 chromosomes (2N = 52). In each cell there are 26 large chromosomes and 26 small chromosomes. Old World cotton plants have 26 chromosomes (2N = 26), all large. Wild American cotton plants have 26 chromosomes, all small. Briefly explain how cultivated American cotton may have originated from Old World cotton and wild American cotton:

_____

_____

_____

6. Discuss the need to maintain the biodiversity of wild plants and ancient farm breeds: _____

_____

_____

_____

The Species Concept

The concept of a species is not as simple as it may first appear. Interbreeding between closely related species, such as the dog family below and 'ring species' on the following page, suggest that the boundaries of a species gene pool can be somewhat unclear. One of the best recognised definitions for a species has been proposed by the respected zoologist, Ernst Mayr: *"A species is a group of actually or potentially interbreeding natural populations that is reproductively isolated from other such groups"*. Each species is provided with a unique classification name to assist with its future identification.

Geographical distribution of selected *Canis* species

The global distribution of most of the species belonging to the genus *Canis* (dogs and wolves) is shown on the map to the right. The **grey wolf** (timber wolf) inhabits the cold, damp forests of North America, northern Europe and Siberia. The range of the three species of **jackal** overlap in the dry, hot, open savannah of Eastern Africa. The now-rare **red wolf** is found only in Texas, while the **coyote** is found inhabiting the open grasslands of the prairies. The **dingo** is found widely distributed throughout the Australian continent inhabiting a variety of habitats. As a result of the spread of human culture, distribution of the domesticated **dog** is global. The dog has been able to interbreed with all other members of the genus listed here, to form fertile hybrids.

Interbreeding between *Canis* species

Members of the genus to which all dogs and wolves belong present problems with the species concept. The domesticated dog is able to breed with numerous other members of the same genus to produce fertile hybrids. The coyote and red wolf in North America have ranges that overlap. They are also able to produce fertile hybrids, although these are rare. By contrast, the ranges of the three distinct species of jackal overlap in the Serengeti of Eastern Africa. These animals are highly territorial, but simply ignore members of the other jackal species and no interbreeding takes place.

For an excellent discussion of species definition among dogs see the article "The Problematic Red Wolf" in *Scientific American*, July 1995, pp. 26-31. This discusses whether or not the red wolf is a species or a long established hybrid of the grey wolf and coyote.

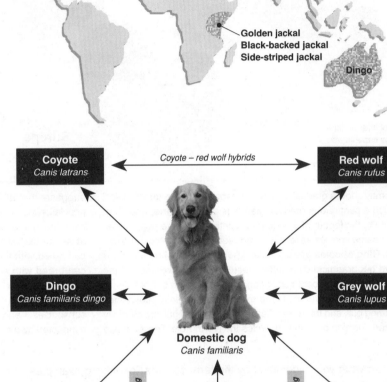

1. Describe the type of barrier that prevents the three species of jackal from interbreeding:

_____

_____

2. Describe the factor that has prevented the dingo from interbreeding with other *Canis* species (apart from the dog):

_____

3. Describe a possible contributing factor to the occurrence of interbreeding between the coyote and red wolf:

_____

_____

4. The grey wolf is a widely distributed species. Explain why the North American population is considered to be part of the same species as the northern European and Siberian populations:

_____

_____

Mechanisms of Evolution

Code: A 2

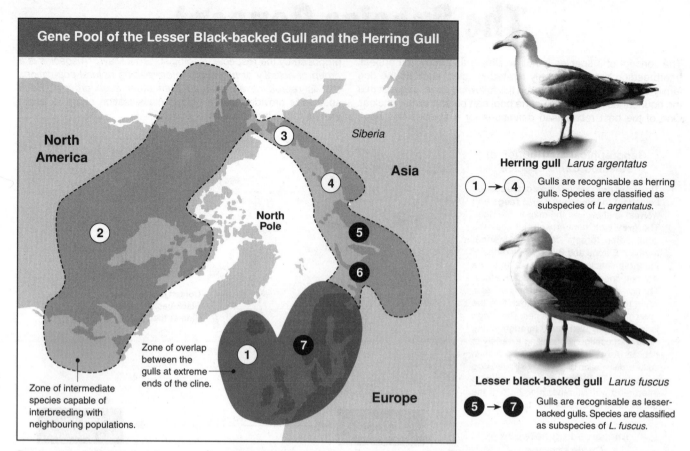

Gene Pool of the Lesser Black-backed Gull and the Herring Gull

North America

Siberia

Asia

North Pole

Europe

Zone of overlap between the gulls at extreme ends of the cline.

Zone of intermediate species capable of interbreeding with neighbouring populations.

Herring gull *Larus argentatus*

1 → 4 Gulls are recognisable as herring gulls. Species are classified as subspecies of *L. argentatus*.

Lesser black-backed gull *Larus fuscus*

5 → 7 Gulls are recognisable as lesser-backed gulls. Species are classified as subspecies of *L. fuscus*.

Species may show a gradual change in phenotype over a geographical area. Such a continuous gradual change is called a **cline**, and often occurs along the length of a country or continent. All the populations are of the same species as long as interbreeding populations link them together. **Ring species** are a special type of cline that has a circular or looped geographical distribution, resulting in the two ends of the cline overlapping. Adjacent populations can interbreed but not where the arms of the loop overlap. In the example above, four subspecies of the herring gull, and three of the lesser black-backed gull are currently recognised, forming a chain that circles the North Pole. The evidence suggests that all subspecies were derived from a single ancestral population that originated in Siberia. Members of this ancestral population migrated in opposite directions, and at the same time evolved so that, at various stages, new subspecies could be identified. Each subspecies can breed with those on either side of it. For instance, subspecies 2 can breed with subspecies 1 and 3, subspecies 4 can breed with subspecies 3 and 5, and so on. However, at the extremes of the cline, which overlaps in northern Europe, the two populations rarely interbreed; subspecies 1 and 7 behave as distinct species even though they are connected by a series of intermediate interbreeding populations.

5. Explain what you understand by the term species, identifying examples where the definition is problematic:

_____

_____

_____

6. The **ring species** (above) do not fit comfortably with the standard definition of a species. Describe the aspects of the population of gulls that:

(a) Supports the idea that they are a single species: _____

_____

_____

_____

_____

(b) Does not agree with the standard definition of a species: _____

_____

_____

_____

_____

Reproductive Isolation

Any factor that prevents two species from producing a fertile hybrid contributes to **reproductive isolation**. Reproductive isolating mechanisms (RIM) are important in preserving the uniqueness of a gene pool. They prevent the dilution effect of **gene flow** into the pool from other populations. Such gene flow may detract from the adaptive combinations already developed as a result of natural selection. A single barrier to reproduction may not completely stop gene flow, so most species have more than one. Geographical barriers are sometimes considered not to be isolating mechanisms because they are not part of the species' biology. Such barriers often precede the development of other isolating mechanisms, which can operate before or after fertilisation. The reproductive isolating mechanisms listed on these pages are separated into two categories: **prezygotic mechanisms**, which operate prior to mating, and **postzygotic mechanisms** that come into play after fertilisation.

Prezygotic Isolating Mechanisms

Spatial (geographical)

Includes physical barriers such as: mountains, rivers, altitude, oceans, isthmuses, deserts, ice sheets. There are many examples of speciation occurring as a result of isolation by oceans or by geological changes in lake basins (e.g. the proliferation of cichlid fish species in Lake Victoria). The many species of iguana from the Galapagos Islands are now quite distinct from the Central and South American species from which they arose.

Temporal (including seasonal)

Timing of mating activity for an organism may prevent contact with closely related species: nocturnal, diurnal, spring, summer, autumn, spring tide etc. Plants flower at different times of the year or even at different times of the day. Closely related animals may have quite different breeding seasons.

Ecological (habitat)

Closely related species may occupy different habitats even when they live in the same general area. Includes small scale differences (e.g. ground dwelling or tree dwelling) and broad differences (forests, grasslands, deserts, freshwater, marine, subterranean, marshes). In New Zealand, Hochstetter's and Archey's frogs occur the same region, but occupy different habitats within that range.

Gamete mortality

Sperm and egg fail to unite. Even if mating takes place, most gametes will fail to unite. The sperm of one species may not be able to survive in the reproductive tract of another species. Gamete recognition may be based on the presence of species specific molecules on the egg or the egg may not release the correct chemical attractants for sperm of another species.

Behavioural (ethological)

Animals attract mates with calls, rituals, dances, body language, etc. Complex displays, such as the flashes of fireflies, are quite specific. In animals, behavioural responses are a major isolating factor, preserving the integrity of mating within species. Birds exhibit a remarkable range of courtship displays that are often quite species-specific.

Structural (morphological)

Shape of the copulatory (mating) apparatus, appearance, coloration, insect attractants. Insects have a lock-and-key arrangement for their copulatory organs. Pheromone chemical attractants, which may travel many kilometres with the aid of the wind, are quite specific, attracting only members of the same species.

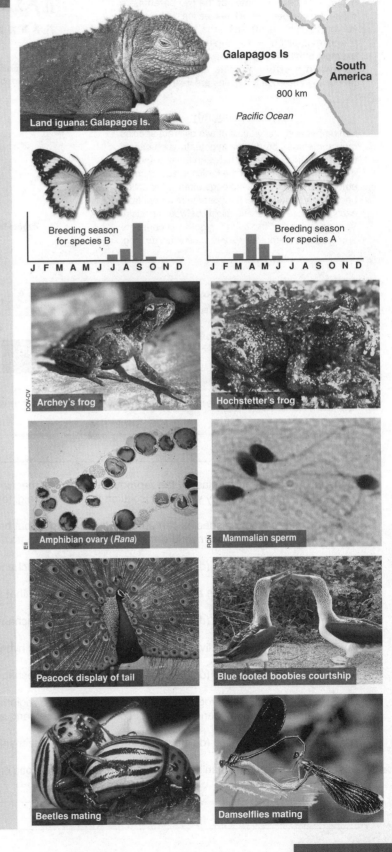

Land iguana: Galapagos Is. Galapagos Is / South America / 800 km / Pacific Ocean

Breeding season for species B — J F M A M J J A S O N D

Breeding season for species A — J F M A M J J A S O N D

Archey's frog / Hochstetter's frog

Amphibian ovary (*Rana*) / Mammalian sperm

Peacock display of tail / Blue footed boobies courtship

Beetles mating / Damselflies mating

Mechanisms of Evolution

© Biozone International 2006
Photocopying Prohibited

Code: A 2

Postzygotic Isolating Mechanisms

Hybrid sterility

Even if two species mate and produce hybrid offspring that are vigorous, the species are still reproductively isolated if the hybrids are sterile (genes cannot flow from one species' gene pool to the other). Such cases are common among the horse family (such as the zebra and donkey shown on the right). One cause of this sterility is the failure of meiosis to produce normal gametes in the hybrid. This can occur if the chromosomes of the two parents are different in number or structure (see the **"zebronkey"** karyotype on the right). The **mule**, a cross between a donkey stallion and a horse mare, is also an example of **hybrid vigour** (they are robust) as well as **hybrid sterility**. Female mules sometimes produce viable eggs but males are infertile.

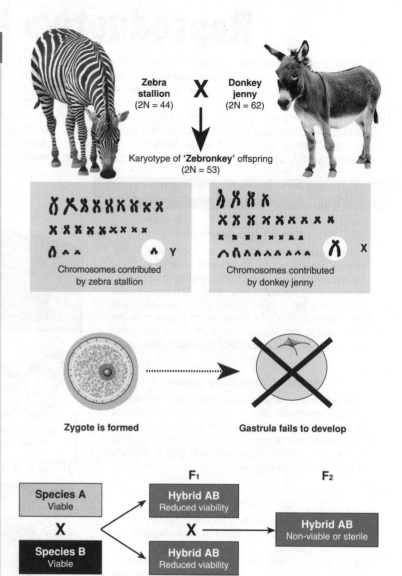

Zebra stallion (2N = 44) X Donkey jenny (2N = 62)

Karyotype of 'Zebronkey' offspring (2N = 53)

Chromosomes contributed by zebra stallion

Chromosomes contributed by donkey jenny

Hybrid inviability

Mating between individuals of two different species may sometimes produce a zygote. In such cases, the genetic incompatibility between the two species may stop development of the fertilised egg at some embryonic stage. Fertilised eggs often fail to divide because of unmatched chromosome numbers from each gamete (a kind of aneuploidy between species). Very occasionally, the hybrid zygote will complete embryonic development but will not survive for long.

Zygote is formed

Gastrula fails to develop

Hybrid breakdown

First generation (F_1) are fertile, but the second generation (F_2) are infertile or inviable. Conflict between the genes of two species sometimes manifests itself in the second generation.

F_1 F_2

Species A — Viable

X

Species B — Viable

Hybrid AB — Reduced viability

Hybrid AB — Reduced viability

X

Hybrid AB — Non-viable or sterile

1. In general terms, explain the role of reproductive isolating mechanisms in maintaining the integrity of a species:

_____

_____

2. In the following examples, classify the reproductive isolating mechanism as either **prezygotic** or **postzygotic** and describe the mechanisms by which the isolation is achieved (e.g. temporal isolation, hybrid sterility etc.):

(a) Some different cotton species can produce fertile hybrids, but breakdown of the hybrid occurs in the next generation when the offspring of the hybrid die in their seeds or grow into defective plants:

Prezygotic / postzygotic (delete one)   Mechanism of isolation: _____

(b) Many plants have unique arrangements of their floral parts that stops transfer of pollen between plants:

Prezygotic / postzygotic (delete one)   Mechanism of isolation: _____

(c) Three species of orchid living in the same rainforest do not hybridise because they flower on different days:

Prezygotic / postzygotic (delete one)   Mechanism of isolation: _____

(d) Several species of the frog genus *Rana*, live in the same regions and habitats, where they may occasionally hybridise. The hybrids generally do not complete development, and those that do are weak and do not survive long:

Prezygotic / postzygotic (delete one)   Mechanism of isolation: _____

3. Postzygotic isolating mechanisms are said to reinforce prezygotic ones. Explain why this is the case:

_____

_____

Allopatric Speciation

Allopatric speciation is a process thought to have been responsible for a great many instances of species formation. It has certainly been important in country which have had a number of cycles of geographical fragmentation. Such cycles can occur as the result of glacial and interglacial periods, where ice expands and then retreats over a land mass. Such events are also accompanied by sea level changes which can isolate populations within relatively small geographical regions.

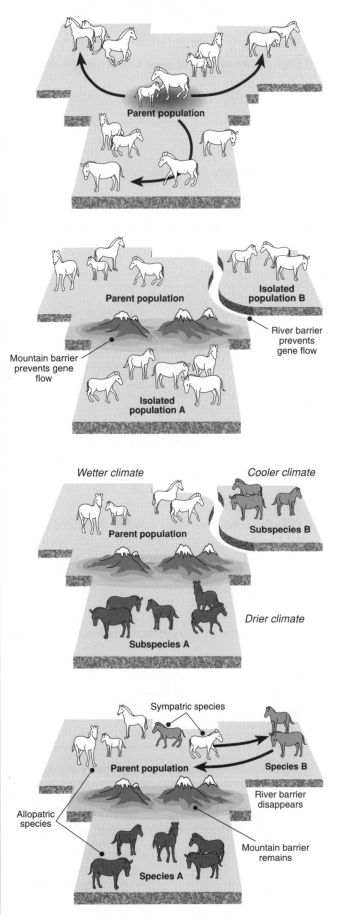

Parent population

Parent population

Isolated population B

River barrier prevents gene flow

Mountain barrier prevents gene flow

Isolated population A

Wetter climate

Cooler climate

Parent population

Subspecies B

Drier climate

Subspecies A

Sympatric species

Parent population

Species B

River barrier disappears

Allopatric species

Mountain barrier remains

Species A

Stage 1: Moving into new environments

There are times when the range of a species expands for a variety of different reasons. A single population in a relatively homogeneous environment will move into new regions of their environment when they are subjected to intense competition (whether it is interspecific or intraspecific). The most severe form of competition is between members of the same species since they are competing for identical resources in the habitat. In the diagram on the right there is a 'parent population' of a single species with a common gene pool with regular 'gene flow' (theoretically any individual has access to all members of the opposite sex for mating purposes).

Stage 2: Geographical isolation

Isolation of parts of the population may occur due to the formation of **physical barriers**. These barriers may cut off those parts of the population that are at the extremes of the species range and gene flow is prevented or rare. The rise and fall of the sea level has been particularly important in functioning as an isolating mechanism. Climatic change can leave 'islands' of habitat separated by large inhospitable zones that the species cannot traverse.

Example: In mountainous regions, alpine species are free to range widely over extensive habitat during cool climatic periods. During warmer periods, however, they may become isolated because their habitat is reduced to 'islands' of high ground surrounded by inhospitable lowland habitat.

Stage 3: Different selection pressures

The isolated populations (A and B) may be subjected to quite different selection pressures. These will favour individuals with traits that suit each particular environment. For example, population A will be subjected to selection pressures that relate to drier conditions. This will favour those individuals with phenotypes (and therefore genotypes) that are better suited to dry conditions. They may for instance have a better ability to conserve water. This would result in improved health, allowing better disease resistance and greater reproductive performance (i.e. more of their offspring survive). Finally, as allele frequencies for certain genes change, the population takes on the status of a **subspecies**. Reproductive isolation is not yet established but the subspecies are significantly different genetically from other related populations.

Stage 4: Reproductive isolation

The separated populations (isolated subspecies) will often undergo changes in their genetic makeup as well as their behaviour patterns. These ensure that the gene pool of each population remains isolated and 'undiluted' by genes from other populations, even if the two populations should be able to remix (due to the removal of the geographical barrier). Gene flow does not occur. The arrows (in the diagram to the right) indicate the zone of overlap between two species after the new Species B has moved back into the range inhabited by the parent population. Closely-related species whose distribution overlaps are said to be **sympatric species**. Those that remain geographically isolated are called **allopatric species**.

Mechanisms of Evolution

Code: RA 2

1. Describe why some animals, given the opportunity, move into new environments: _____

_____

_____

2. (a) Plants are unable to move. Explain how plants might disperse to new environments: _____

_____

   (b) Describe the amount of **gene flow** within the parent population prior to and during this range expansion:

_____

_____

_____

3. Identify the **process** that causes the formation of new **mountain ranges**: _____

4. Identify the event that can cause large changes in **sea level** (up to 200 metres): _____

5. Describe six **physical barriers** that could isolate different parts of the same population: _____

   (a) _____   (d) _____

   (b) _____   (e) _____

   (c) _____   (f) _____

6. Describe the effect that physical barriers have on **gene flow**: _____

_____

_____

_____

7. (a) Describe four different types of **selection pressure** that could have an effect on a gene pool: _____

_____

_____

   (b) Describe briefly how these selection pressures affect the isolated gene pool in terms of **allele frequencies**:

_____

_____

_____

8. Describe two types of **prezygotic** and two types of **postzygotic** reproductive isolating mechanisms (see previous pages):

   (a) Prezygotic: _____

_____

   (b) Postzygotic: _____

_____

9. Distinguish between **allopatry** and **sympatry** in populations: _____

_____

_____

_____

_____

Sympatric Speciation

New species may be formed even where there is no separation of the gene pools by physical barriers. Called **sympatric speciation**, it is rarer than allopatric speciation, although not uncommon in plants which form **polyploids**. There are two situations where sympatric speciation is thought to occur. These are described below:

Speciation Through Niche Differentiation

Niche isolation

In a heterogeneous environment (one that is not the same everywhere), a population exists within a diverse collection of **microhabitats**. Some organisms prefer to occupy one particular type of 'microhabitat' most of the time, only rarely coming in contact with fellow organisms that prefer other microhabitats. Some organisms become so dependent on the resources offered by their particular microhabitat that they never meet up with their counterparts in different microhabitats.

Reproductive isolation

Finally, the individual groups have remained genetically isolated for so long because of their microhabitat preferences, that they have become reproductively isolated. They have become new species that have developed subtle differences in behaviour, structure, and physiology. Gene flow (via sexual reproduction) is limited to organisms that share a similar microhabitat preference (as shown in the diagram on the right).

Example: Some beetles prefer to find plants identical to the species they grew up on, when it is time for them to lay eggs. Individual beetles of the same species have different preferences.

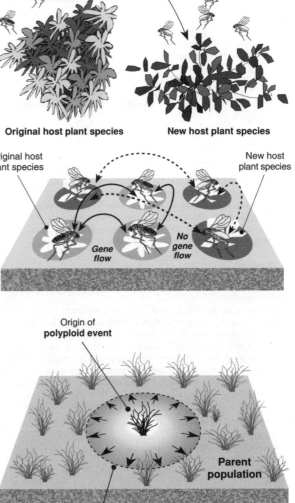

An insect forced to lay its eggs on an unfamiliar plant species may give rise to a new population of flies isolated from the original population.

Original host plant species   **New host plant species**

Original host plant species   New host plant species

Gene flow   **No gene flow**

Instant Speciation by Polyploidy

Polyploidy may result in the formation of a new species without isolation from the parent species. This event, occurring during meiosis, produces sudden reproductive isolation for the new group. Because the sex-determining mechanism is disturbed, animals are rarely able to achieve new species status this way (they are effectively sterile, e.g. tetraploid XXXX). Many plants, on the other hand, are able to reproduce vegetatively, or carry out self pollination. This ability to reproduce on their own enables such polyploid plants to produce a breeding population.

Speciation by allopolyploidy

This type of polyploidy usually arises from the doubling of chromosomes in a hybrid between two different species. The doubling often makes the hybrid fertile.

Examples: Modern wheat. Swedes are polyploid species formed from a hybrid between a type of cabbage and a type of turnip.

Origin of **polyploid event**

Parent population

New polyploid plant species spreads outwards through the existing parent population

1. Explain what is meant by **sympatric speciation** (do not confuse this with sympatric species): _____

_____

_____

2. Explain briefly how polyploidy may cause the formation of a new species: _____

_____

_____

3. Identify an example of a species that has been formed by polyploidy: _____

4. Explain briefly how niche differentiation may cause the formation of a new species: _____

_____

_____

Mechanisms of Evolution

Code: A 2

Evolution in Bacteria

As a result of their short **generation times**, bacterial populations can show significant evolutionary change within relatively short periods of time. The development of **antibiotic resistance** is one such evolutionary change and it arises and spreads within and between bacterial populations with frightening ease. A variety of human practices have led to antibiotic resistance and have increased the rate at which bacterial strains acquire new properties.

These practices include the overuse and misuse of antibiotics by physicians, the use of antibiotics by immunosuppressed patients to prevent infection, the use of antibiotics in animal feed, and the spread of resistant bacteria to new areas because of air travel. For many strains of pathogenic bacteria, resistant mutants are increasingly replacing susceptible normal populations. This makes the search for new types of antibiotics increasingly urgent.

The Evolution of Drug Resistance in Bacteria

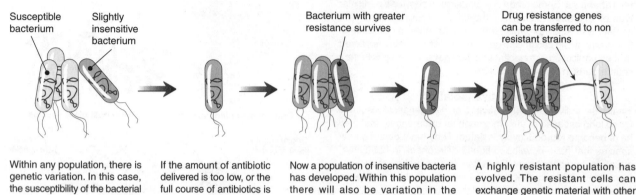

Susceptible bacterium | Slightly insensitive bacterium | Bacterium with greater resistance survives | Drug resistance genes can be transferred to non resistant strains

Within any population, there is genetic variation. In this case, the susceptibility of the bacterial strain is normally distributed, with some cells being more susceptible than others.

If the amount of antibiotic delivered is too low, or the full course of antibiotics is not completed, only the most susceptible bacteria will die.

Now a population of insensitive bacteria has developed. Within this population there will also be variation in the susceptibility to antibiotics. As treatment continues, some of the bacteria may acquire greater resistance.

A highly resistant population has evolved. The resistant cells can exchange genetic material with other bacteria, passing on the resistance genes. The antibiotic that was initially used against this bacterial strain will now be ineffective against it.

Observing Adaptive Radiation

Recently, scientists have demonstrated rapid evolution in bacteria. *Pseudomonas fluorescens* was used in the experiment and propagated in a simple heterogeneous environment consisting of a 25 cm³ glass container containing 6 cm³ of broth medium. Over a short period of time, the bacteria underwent morphological diversification, with a number of new morphs appearing. These morphs were shown to be genetically distinct. A striking feature of the evolved species is their

niche specificity, with each new morph occupying a distinct habitat (below, left). In a follow up experiment (below, right), the researchers grew the same original bacterial strain in the same broth under identical incubation conditions, but in a homogeneous environment (achieved by shaking the broth). Without the different habitats offered by an undisturbed environment, no morphs emerged. The experiment illustrated the capacity of bacteria to evolve to utilise available niches.

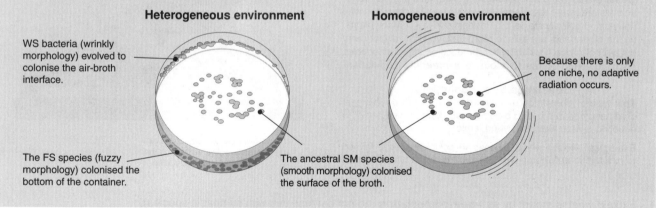

Heterogeneous environment

WS bacteria (wrinkly morphology) evolved to colonise the air-broth interface.

The FS species (fuzzy morphology) colonised the bottom of the container.

Homogeneous environment

Because there is only one niche, no adaptive radiation occurs.

The ancestral SM species (smooth morphology) colonised the surface of the broth.

1. Using an illustrative example, explain why evolution of new properties in bacteria can be very rapid:

_____

_____

2. (a) In the example above, explain why the bacteria evolved when grown in a heterogeneous environment:

_____

_____

(b) Predict what would happen if the FS morph was cultured in the homogeneous environment: _____

_____

_____

Code: A 2

Stages in Species Development

The diagram below represents a possible sequence of genetic events involved in the origin of two new species from an ancestral population. As time progresses (from top to bottom of the diagram) the amount of genetic variation increases and each group becomes increasingly isolated from the other. The mechanisms that operate to keep the two gene pools isolated from one another may begin with **geographical barriers**. This may be followed by **prezygotic** mechanisms which protect the gene pool from unwanted dilution by genes from other pools. A longer period of isolation may lead to **postzygotic** mechanisms (see the page on reproductive isolating mechanisms). As the two gene pools become increasingly isolated and different from each other, they are progressively labelled: population, race, and subspecies. Finally they attain the status of separate species.

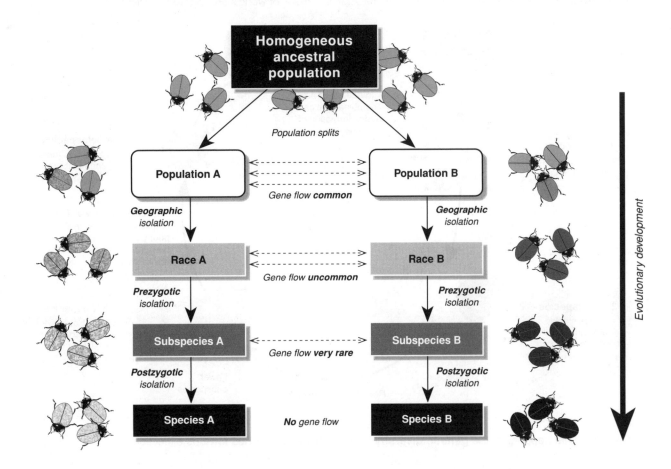

1. Explain what happens to the extent of gene flow between diverging populations as they gradually attain species status:

_____

_____

2. Early human populations about 500 000 ya were scattered across Africa, Europe, and Asia. This was a time of many regional variants, collectively called archaic *Homo sapiens*. The fossil skulls from different regions showed mixtures of characteristics, some modern and some 'primitive'. These regional populations are generally given subspecies status. Suggest reasons why gene flow between these populations may have been rare, but still occasionally occurred:

_____

_____

_____

3. In the USA, the species status of several duck species, including the black duck (*Anas rubripes*) and the mottled duck in Florida (*A. fulvigula*) is threatened by interbreeding with the now widespread and very adaptable mallard duck (*A. platyrhynchos*). Similar threatened extinction though hybridisation has occurred in New Zealand, where the native grey duck has been virtually eliminated as a result of interbreeding with the introduced mallard.

   (a) Suggest why these hybrids threaten the Species status of some native duck species: _____

   _____

   (b) Suggest what factor may deter mallards from hybridising with other duck species: _____

   _____

Mechanisms of Evolution

Code: A 3

The Species Life Cycle

Species have a 'life cycle' that is different to that of an individual organism. In the early stages of the emergence of a new species, it is often in a position to exploit new habitats and niches in its search to escape the ever present forces of competition and exploitation. This leads to the species increasing its range as the population grows in numbers. An increased diversity in the gene pool develops as the population begins to respond to the natural selection pressures of each of the new habitat variations. At this stage, there may exist merely 'racial' types or even subspecies (groups that have significant genotypic and phenotypic variations, but whose reproductive isolation is not complete and is largely controlled by physical barriers). If a species begins to decline in its ability to successfully cope with a changing environment, then the number of subspecies reduces and finally the numbers and range of individuals can reduce to the point of extinction (e.g. many of world's endangered species).

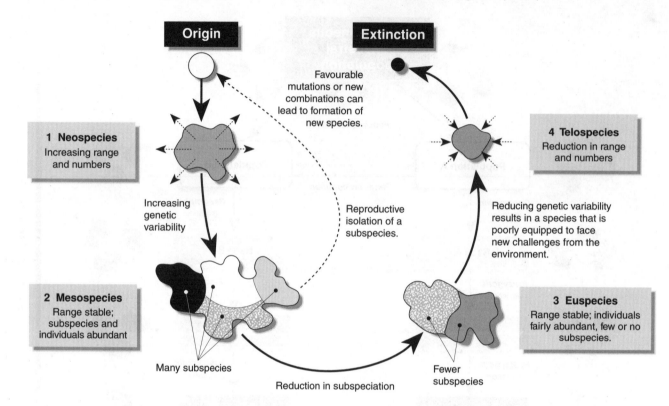

1. Describe what happens to the **genetic diversity** of a species as it declines towards extinction: _____

_____

_____

2. Describe two situations where a species may be able to exploit new opportunities and become more 'successful':

    (a) _____

    _____

    (b) _____

    _____

3. Bearing in mind what is happening to many of today's endangered species, describe four reasons, **not** due to human influences, why a species may decline to the point of extinction:

    (a) _____

    (b) _____

    (c) _____

    (d) _____

4. Describe two examples where humans have reduced the genetic diversity of other species:

    (a) _____

    (b) _____

Patterns of Evolution

Describing processes and patterns of evolution

Convergent, divergent, and parallel evolution, coevolution, examples of adaptive radiation, the rate of evolutionary change, and extinction

Learning Objectives

☐ 1. Compile your own glossary from the **KEY WORDS** displayed in **bold type** in the learning objectives below.

Patterns of Evolution *(pages 74, 76-88)*

☐ 2. Using examples, distinguish patterns of species formation: **sequential (phyletic) speciation, coevolution, divergent evolution, adaptive radiation** (dichotomous) speciation. Recognise adaptive radiation as a form of **divergent evolution**. Explain how evolutionary change over time has resulted in a great diversity of forms among living organisms.

☐ 3. Explain **convergent evolution** and provide examples. Discuss how **analogous structures** (analogies) may arise as a result of convergence. Distinguish clearly between **analogous structures** and **homologous structures** and explain the role of homology in identifying evolutionary relationships.

☐ 4. Describe examples of **coevolution**, including in flowering plants and their pollinators, parasites and their hosts, and predators and their prey (including herbivory). Discuss the evidence for coevolution in species with close ecological relationships.

☐ 5. Understand that some biologists also recognise **parallel evolution** to indicate evolution along similar lines in closely related groups.

☐ 6. Distinguish between the **punctuated equilibrium** and **gradualism** models for the pace of evolutionary change. Discuss the evidence for each model and discuss the evidence for each in different taxa.

☐ 7. Explain what is meant by **extinction**, identifying it as part of of the **species life cycle**. Describe the role of extinction in evolution. Distinguish clearly between **background extinction rates** and **mass extinction**. Identify the major **mass extinctions** and discuss the theories for their causes.

☐ 8. Describe examples of evolution (including speciation). For each example, describe:
  (a) The genus, species, and sub-species involved.
  (b) Important features of the species divergence:
   • Geographical barriers between populations.
   • Habitat range and niche differentiation.
   • Any zones of overlap in distribution (sympatry).
   • Recent range expansions.
  (c) Evidence for the evolution, including genetic studies.

Classification and Phylogeny *(pages 89-99)*

☐ 9. Recall the **binomial system** for the classification of organisms. Describe characteristics of the kingdoms, and the major animal phyla and plant divisions.

☐ 10. Explain how classification systems (should) reflect the evolutionary relationships and history (**phylogeny**) of organisms. Describe the evolution and classification of a taxonomic group.

See page 7 for additional details of this text:
■ Futuyma, D.J., 2005. **Evolution**, (Sinauer Associates), chpt. 18-21 as required.
■ Martin, R.A., 2004. **Missing Links**, chpt. 2 and case histories from section II as required.
■ Zimmer, C., 2001. **Evolution: The Triumph of an Idea**, (HarperCollins), chpt. 1-4, 6 as required.

See page 7 for details of publishers of periodicals:

STUDENT'S REFERENCE

■ **Dinosaurs take Wing** National Geographic, 194(1) July 1998, pp. 74-99. *An account of the evolution of birds from small theropod dinosaurs, including an exploration of the homology between the typical dinosaur limb and the wing of the modern bird. An excellent account.*

■ **The Rise of Mammals** National Geographic, 203(4), April 2003, pp. 2-37. *An account of the adaptive radiation of mammals and the significance of the placenta in mammalian evolution.*

■ **Together We're Stronger** New Scientist, 15 March 2003. (Inside Science). *The mechanisms behind the evolution of social behaviour in animals.*

■ **Mass Extinctions** New Scientist, 11 Dec. 1999 (Inside Science). *The nature and possible causes of the five mass extinctions of the past. includes discussion of the current sixth extinction.*

■ **The Sixth Extinction** National Geographic, 195(2) Feb. 1999, pp. 42-59. *High extinction rates have occurred five times in the past. Human impact is driving the current sixth extinction.*

■ **Evolution: Five Big Questions** New Scientist, 14 June 2003, pp. 32-39, 48-51. *A discussion of the most covered points regarding evolution and the mechanisms by which it occurs.*

■ **Which Came First?** Scientific American, Feb. 1997, pp. 12-14. *Shared features among fossils; the result from convergence or common ancestry?*

■ **In the Blink of an Eye** New Scientist, 9 July 2005, pp. 28-31. *Rapid contemporary evolution may be widespread and humans may be unwittingly helping it along.*

■ **Resources for Teaching Evolution** The American Biology Teacher, 66(2), Feb. 2004, pp. 109-113. *The latest approaches to teaching evolution, with reference to sample activities.*

See pages 4-5 for details of how to access **Bio Links** from our web site: **www.thebiozone.com** From Bio Links, access sites under the topics:

GENERAL BIOLOGY ONLINE RESOURCES > Online Textbooks and Lecture Notes: • An on-line biology book... *and others* > **General Online Biology resources:** • Ken's Bioweb resources ... *and others* > **Glossaries:** • Evolutionary biology and genetics glossary... *and others*

  EVOLUTION: • A history of evolutionary thought • Evolution for teaching • BIO 414 evolution • Enter evolution: theory and history • Evolution • Evolution on the web for biology students • Harvard University biology links: evolution • The Talk.Origins archive • Whale origins ... *and others*

Presentation MEDIA to support this topic:

EVOLUTION

Patterns of Evolution

The diversification of an ancestral group into two or more species in different habitats is called **divergent evolution**. This process is illustrated in the diagram below, where two species have diverged from a **common ancestor**. Note that another species budded off, only to become extinct. Divergence is common in evolution. When divergent evolution involves the formation of a large number of species to occupy different niches, this is called an **adaptive**

radiation. The example below (right) describes the radiation of the mammals that occurred after the extinction of the dinosaurs; an event that made niches available for exploitation. Note that the evolution of species may not necessarily involve branching; a species may accumulate genetic changes that, over time, result in the emergence of what can be recognised as a different species. This is known as **sequential evolution** (below, left).

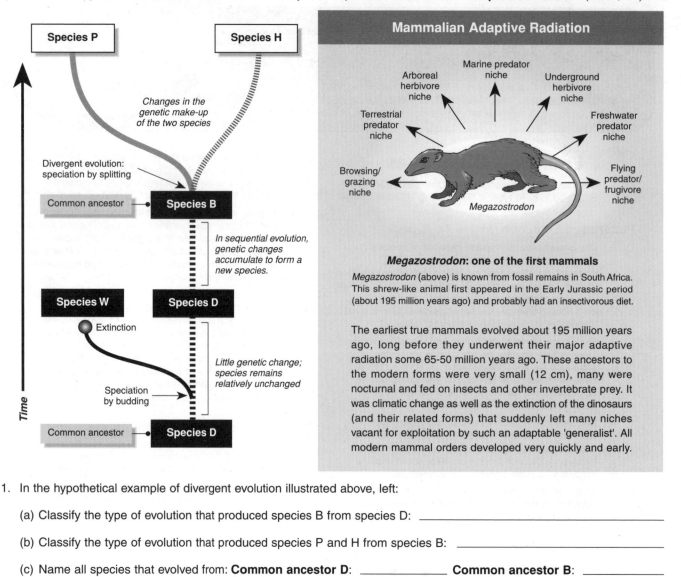

1. In the hypothetical example of divergent evolution illustrated above, left:

   (a) Classify the type of evolution that produced species B from species D: _____

   (b) Classify the type of evolution that produced species P and H from species B: _____

   (c) Name all species that evolved from: **Common ancestor D**: _____ **Common ancestor B**: _____

   (d) Suggest why species B, P, and H all possess a physical trait not found in species D or W: _____

   _____

2. (a) Explain the distinction between **divergence** and **adaptive radiation**: _____

   _____

   _____

   (b) Discuss the differences between **sequential evolution** and **divergent evolution**: _____

   _____

   _____

   _____

   _____

   _____

The Rate of Evolutionary Change

The pace of evolution has been much debated, with two models being proposed: **gradualism** and **punctuated equilibrium**. Some scientists believe that both mechanisms may operate at different times and in different circumstances. Interpretations of the fossil record will vary depending on the time scales involved. During its formative millenia, a species may have accumulated its changes gradually (e.g. over 50 000 years). If that species survives for 5 million years, the evolution of its defining characteristics would have been compressed into just 1% of its (species) lifetime. In the fossil record, the species would appear quite suddenly.

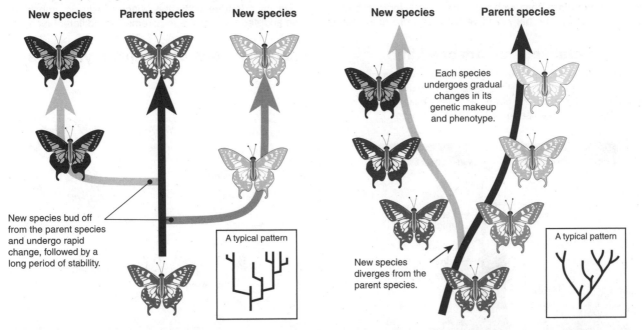

New species bud off from the parent species and undergo rapid change, followed by a long period of stability.

A typical pattern

Each species undergoes gradual changes in its genetic makeup and phenotype.

New species diverges from the parent species.

A typical pattern

Punctuated Equilibrium

There is abundant evidence in the fossil record that, instead of gradual change, species stayed much the same for long periods of time (called stasis). These periods were punctuated by short bursts of evolution which produce new species quite rapidly. According to the punctuated equilibrium theory, most of a species' existence is spent in stasis and little time is spent in active evolutionary change. The stimulus for evolution occurs when some crucial factor in the environment changes.

Gradualism

Gradualism assumes that populations slowly diverge from one another by accumulating adaptive characteristics in response to different selective pressures. If species evolve by gradualism, there should be transitional forms seen in the fossil record, as is seen with the evolution of the horse. Trilobites, an extinct marine arthropod, are another group of animals that have exhibited gradualism. In a study in 1987 a researcher found that they changed gradually over a three million year period.

1. Discuss the nature of the environments that would support each of the following paces of evolutionary change:

    (a) Punctuated equilibrium: _____

    _____

    _____

    (b) Gradualism: _____

    _____

    _____

2. In the fossil record of early human evolution, species tend to appear suddenly, linger for often very extended periods before disappearing suddenly. There are few examples of smooth inter-gradations from one species to the next. Explain which of the above models best describes the rate of human evolution:

    _____

3. Some species apparently show little evolutionary change over long periods of time (hundreds of millions of years).

    (a) Name two examples of such species: _____

    (b) State the term given to this lack of evolutionary change: _____

    (c) Explain why such species have changed little over evolutionary time: _____

    _____

Code: RA 2

Homologous Structures

The evolutionary relationships between groups of organisms is determined mainly by structural similarities called **homologous structures** (homologies), which suggest that they all descended from a common ancestor with that feature. The bones of the forelimb of air-breathing vertebrates are composed of similar bones arranged in a comparable pattern. This is indicative of a common ancestry. The early land vertebrates were amphibians and possessed a limb structure called the **pentadactyl limb**: a limb with five fingers or toes (below left). All vertebrates that descended from these early amphibians, including reptiles, birds and mammals, have limbs that have evolved from this same basic pentadactyl pattern. They also illustrate the phenomenon known as **adaptive radiation**, since the basic limb plan has been adapted to meet the requirements of different niches.

Generalised Pentadactyl Limb

The forelimbs and hind limbs have the same arrangement of bones but they have different names. In many cases bones in different parts of the limb have been highly modified to give it a specialised locomotory function.

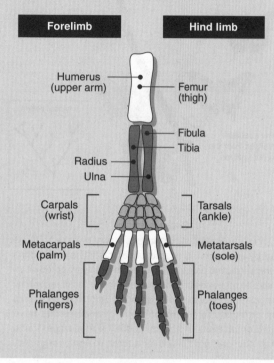

Forelimb	Hind limb

Humerus (upper arm) — Femur (thigh)

Fibula
Tibia

Radius
Ulna

Carpals (wrist) — Tarsals (ankle)

Metacarpals (palm) — Metatarsals (sole)

Phalanges (fingers) — Phalanges (toes)

Specialisations of Pentadactyl Limbs

Bird's wing

Mole's forelimb

Bat's wing

Dog's front leg

Seal's flipper

Human arm

1. Briefly describe the purpose of the major anatomical change that has taken place in each of the limb examples above:

   (a) Bird wing: *Highly modified for flight. Forelimb is shaped for aerodynamic lift and feather attachment.*

   (b) Human arm: _____

   (c) Seal flipper: _____

   (d) Dog foot: _____

   (e) Mole forelimb: _____

   (f) Bat wing: _____

2. Describe how homology in the pentadactyl limb is evidence for adaptive radiation: _____

   _____

   _____

   _____

3. Homology in the behaviour of animals (for example, sharing similar courtship or nesting rituals) is sometimes used to indicate the degree of relatedness between groups. Suggest how behaviour could be used in this way:

   _____

   _____

   _____

Convergent Evolution

Not all similarities between species are inherited from a common ancestor. Species from different evolutionary branches may come to resemble each other if they have similar ecological roles and natural selection has shaped similar adaptations. This is called **convergent evolution** or **convergence**. Similarity of form due to convergence is called **analogy**.

Convergence in Swimming Form

Although similarities in body form and function can arise because of common ancestry, it may also be a result of **convergent evolution**. Selection pressures in a particular environment may bring about similar adaptations in unrelated species. These selection pressures require the solving of problems in particular ways, leading to the similarity of body form or function. The development of succulent forms in unrelated plant groups (*Euphorbia* and the cactus family) is an example of convergence in plants. In the example (right), the selection pressures of the aquatic environment have produced a similar **streamlined** body shape in unrelated vertebrate groups. Icthyosaurs, penguins, and dolphins each evolved from terrestrial species that took up an aquatic lifestyle. Their general body form has evolved to become similar to that of the shark, which has always been aquatic. Note that flipper shape in mammals, birds, and reptiles is a result of convergence, but its origin from the pentadactyl limb is an example of **homology**.

Analogous Structures

Analogous structures are those that have the same function and often the same basic external appearance, but **quite different origins**. The example on the right illustrates how a complex eye structure has developed independently in two unrelated groups. The appearance of the **eye** is similar, but there is no genetic relatedness between the two groups (mammals and cephalopod molluscs). The **wings** of birds and insects are also an example of analogy. The wings perform the same function, but the two groups share no common ancestor. *Longisquama*, a lizard-like creature that lived about 220 million years ago, also had 'wings' that probably allowed gliding between trees. These 'wings' were not a modification of the forearm (as in birds), but highly modified long scales or feathers extending from its back.

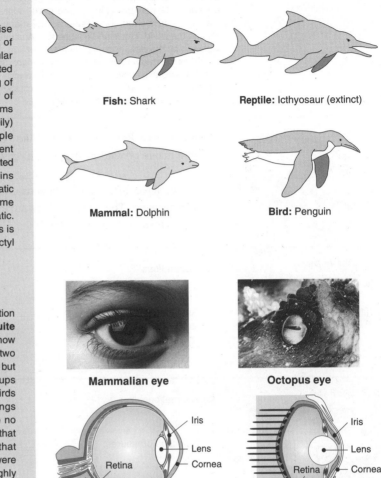

Fish: Shark

Reptile: Icthyosaur (extinct)

Mammal: Dolphin

Bird: Penguin

Mammalian eye

Octopus eye

Iris

Lens

Retina

Cornea

Iris

Lens

Retina

Cornea

1. In the example above illustrating convergence in swimming form, describe two ways in which the body form has evolved in response to the particular selection pressures of the aquatic environment:

   (a) _____

   _____

   (b) _____

   _____

2. Describe two of the selection pressures that have influenced the body form of the swimming animals above:

   (a) _____

   (b) _____

3. When early taxonomists encountered new species in the Pacific region and the Americas, they were keen to assign them to existing taxonomic families based on their apparent similarity to European species. In recent times, many of the new species have been found to be quite unrelated to the European families they were assigned to. Explain why the traditional approach did not reveal the true evolutionary relationships of the new species:

   _____

   _____

   _____

   _____

80

4. For each of the paired examples, briefly describe the adaptations of body shape, diet and locomotion that appear to be similar in both forms, and the likely selection pressures that are acting on these mammals to produce similar body forms:

Convergence Between Marsupials and Placentals

Marsupials and placental mammals were separated from each other very early in mammalian evolution (about 120 mya). Marsupials were initially widely distributed throughout the ancient supercontinent of Gondwana, and there are some modern species still living in the American continent. Gondwana split up about 100 million years ago. As the placentals developed, they displaced the marsupials in most habitats around the world. The island continent of Australia, because of its early isolation by the sea, escaped this competition and placentals did not reach the continent until the arrival of humans 35 000 to 50 000 years ago. The Australian marsupials evolved into a wide variety of forms (below left) that bear a remarkable resemblance to ecologically equivalent species of North American placentals (below right).

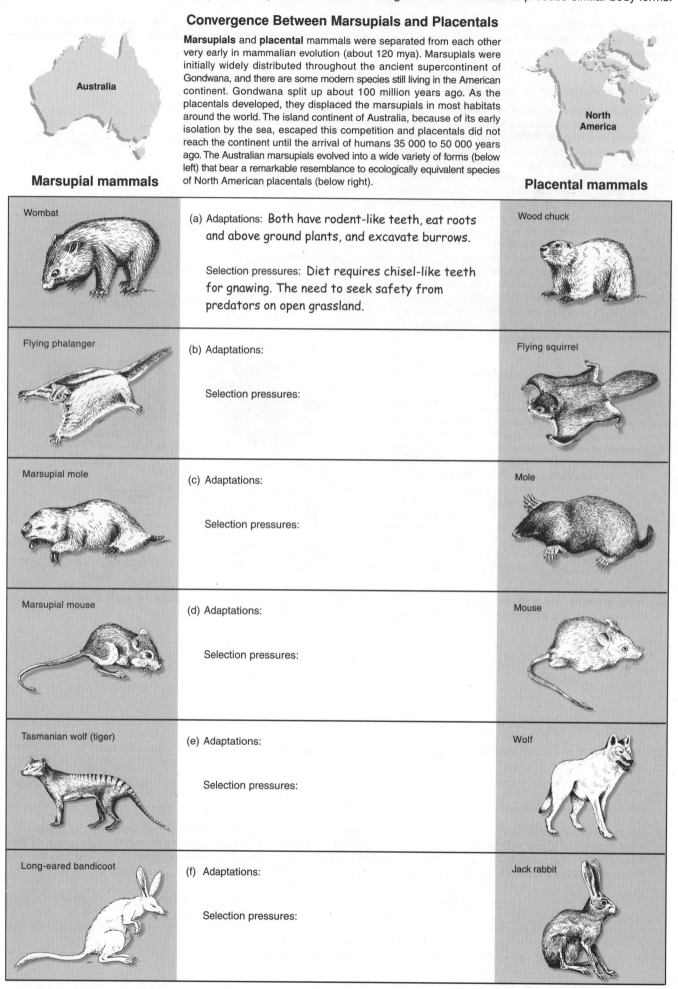

Marsupial mammals | **Placental mammals**

Wombat / Wood chuck

(a) Adaptations: Both have rodent-like teeth, eat roots and above ground plants, and excavate burrows.

Selection pressures: Diet requires chisel-like teeth for gnawing. The need to seek safety from predators on open grassland.

Flying phalanger / Flying squirrel

(b) Adaptations:

Selection pressures:

Marsupial mole / Mole

(c) Adaptations:

Selection pressures:

Marsupial mouse / Mouse

(d) Adaptations:

Selection pressures:

Tasmanian wolf (tiger) / Wolf

(e) Adaptations:

Selection pressures:

Long-eared bandicoot / Jack rabbit

(f) Adaptations:

Selection pressures:

Coevolution

The term **coevolution** is used to describe cases where two (or more) species reciprocally affect each other's evolution. Each party in a coevolutionary relationship exerts selective pressures on the other and, over time, the species develop a relationship that may involve mutual dependency. Coevolution is a likely consequence when different species have close ecological interactions with one another. These ecological relationships include predator-prey and parasite-host relationships and mutualistic relationships such as those between plants and their pollinators (see *Pollination Syndromes*). There are many examples of coevolution amongst parasites or pathogens and their hosts, and between predators and their prey, as shown on the following page.

Photo courtesy of Alex Wild

Swollen-thorn *Acacia* lack the cyanogenic glycosides found in related *Acacia* spp. and the thorns are large and hollow, providing living space for the aggressive, stinging *Pseudomyrmex* ants which patrol the plant and protect it from browsing herbivores. The *Acacia* also provides the ants with protein rich food.

Hummingbirds (above) are important pollinators in the tropics. Their needle-like bills and long tongues can take nectar from flowers with deep tubes. Their ability to hover enables them to feed quickly from dangling flowers. As they feed, their heads are dusted with pollen, which is efficiently transferred between flowers.

Butterflies find flowers by vision and smell them after landing to judge their nectar source. Like bees, they can remember characteristics of desirable flowers and so exhibit constancy, which benefits both pollinator and plant. Butterfly flowers are very fragrant and are blue, purple, deep pink, red, or orange.

Bees are excellent pollinators; they are strong enough to enter intricate flowers and have medium length tongues which can collect nectar from many flower types. They have good colour vision, which extends into the UV, but they are red-blind, so bee pollinated flowers are typically blue, purplish, or white and they may have nectar guides that are visible as spots.

Beetles represent a very ancient group of insects with thousands of modern species. Their high diversity has been attributed to extensive coevolution with flowering plants. Beetles consume the ovules as well as pollen and nectar and there is evidence that ovule herbivory by beetles might have driven the evolution of protective carpels in angiosperms.

NZ's short tailed bat pollinates *Dactylanthus* flowers on the forest floor

DoC

Bats are nocturnal and colour-blind but have an excellent sense of smell and are capable of long flights. Flowers that have coevolved with bat pollinators are open at night and have light or drab colours that do not attract other pollinators. Bat pollinated flowers also produce strong fragrances that mimic the smell of bats and have a wide bell shape for easy access.

1. Using examples, explain what you understand by the term coevolution: _____

_____

_____

_____

_____

2. Describe some of the strategies that have evolved in plants to attract pollinators: _____

_____

_____

_____

_____

_____

Code: RA 3

Predators, Parasites, and Coevolution

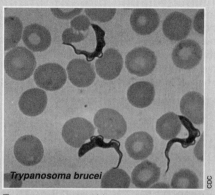

Trypanosoma brucei

Predators have obviously evolved to exploit their prey, with effective offensive weapons and hunting ability being paramount. Prey have evolved numerous strategies to protect themselves from predators, including large size and strength, protective coverings, defensive weapons, and toxicity. Lions have evolved the ability to hunt cooperatively to increase their chance of securing a kill from swift herding species such as zebra and gazelles.

Female *Helicornius* butterflies will avoid laying their eggs on plants already occupied by eggs, because their larvae are highly cannibalistic. Passionfruit plants (*Passiflora*) have exploited this by creating fake, yellow eggs on leaves and buds. *Passiflora* has many chemical defences against herbivory, but these have been breached by *Heliconius*. It has thus counter-evolved new defences against this herbivory by this genus.

Trypanosomes provide a good example of **host-parasite coevolution**. Trypanosomes must evolve strategies to evade their host's defences, but their virulence is constrained by needing to keep their host alive. Molecular studies show that *Trypanosoma brucei* coevolved in Africa with the first hominids around 5 mya, but *T. cruzi* contact with human hosts occurred in South America only after settlements were made by nomadic cultures.

3. Explain how coevolution could lead to an increase in biodiversity: _____

_____

_____

_____

_____

4. Discuss some of the possible consequences of species competition: _____

_____

_____

_____

_____

_____

_____

_____

5. The analogy of an "arms race" is often used to explain the coevolution of exploitative relationships such as those of a parasite and its host. Form a small group to discuss this idea and then suggest how the analogy is flawed:

_____

_____

_____

_____

_____

_____

_____

_____

Pollination Syndromes

The mutualistic relationship between plants and their pollinators represents a classic case of coevolution. Flower structure has evolved in many different ways in response to the many types of animal pollinators. Flowers and pollinators have coordinated traits known as **pollination syndromes**. This makes it relatively easy to deduce pollinators type from the appearance of flowers (and vice versa). Plants and animals involved in such pollination associations often become highly specialised in ways that improve pollination efficiency: innovation by one party leads to some response from the other.

Controlling Pollinator Access

Flowers control pollinator access by flower shape and position.

Dandelion

Rigid inflorescences offer a stable landing platform to small or heavy insects, such as bumblebees.

Fuschia

Only animals that can hover can collect rewards from and pollinate flowers that hang upside down.

Attracting Pollinators

Flowers advertise the presence of nectar and pollen, with colour, scent, shape, and arrangement.

Rose

Daisy

Nectar guides help the pollinator to locate nectar and pollen. In this flower, the inner petals reflect UV.

While many flowers, like roses, are fragrant, flowers pollinated by flies (right) can give off dung or rotten meat smells.

Common Pollination Syndromes: Insects

Beetles

Ancient insect group
Good sense of smell
Hard, smooth bodies

Beetle-pollinated flowers

Ancient plant groups
Strong, fruity odours
Large, often flat, with easy access

Nectar-feeding flies

Sense nectar with feet
Tubular mouthparts

Nectar-feeding fly-pollinated flowers

Simple flowers with easy access
red or light colour, little odour

Moths

Many active at night
Good sense of smell
Feed with long, narrow tongues
Some need landing platforms

Moth-pollinated flowers

Flowers may be open at night
Fragrant; with heavy, musky scent
Nectar in narrow, deep tubes
landing platforms often provided

Carrion flies

Attracted by heat, odours, or
or colour of carrion or dung.
Food in the form of nectar or
pollen not required.

Carrion fly-pollinated flowers

Coloured to resemble dung or carrion
Produce heat or foul odours
No nectar or pollen reward offered

Common Pollination Syndromes: Vertebrates

Birds

Most require a perching site
Good colour vision, including red
Poor sense of smell
Feed during daylight
High energy requirements

Bird-pollinated flowers

Large and damage resistant
Often red or other bright colours
Not particularly fragrant
Open during the day
Copious nectar produced

Bats

Active at night
High food requirements
Colour blind
Good sense of smell
Cannot fly in foliage
High blossom intelligence

Bat-pollinated flowers

Open at night
Plentiful nectar and pollen offered
Light or dingy colours
Strong, often bat-like odours
Open shape, easy access
Pendulous or on the trunks of trees

Non-flying mammals

Relatively large size
High energy requirements
Colour vision may be lacking
Good sense of smell

Non-bat mammal-pollinated flowers

Robust, damage resistant
Copious, sugar-rich nectar
Dull coloured
Odorous, but not necessarily fragrant

1. (a) Describe a common pollination syndrome of an insect: _____

    _____

   (b) Describe a common pollination syndrome of a vertebrate: _____

    _____

2. Suggest how knowledge of pollination syndromes might be used to develop testable predictions about plant and animal pollination relationships:

    _____

    _____

    _____

    _____

Code: A 2

Geographical Distribution

The camel family, Camelidae, consists of six modern-day species that have survived on three continents: Asia, Africa and South America. They are characterised by having only two functional toes, supported by expanded pads for walking on sand or snow. The slender snout bears a cleft upper lip. The recent distribution of the camel family is fragmented. Geophysical forces such as plate tectonics and the ice age cycles have controlled the extent of their distribution. South America, for example, was separated from North America until the end of the Pliocene, about 2 million years ago. Three general principles about the dispersal and distribution of land animals are:

- When very closely related animals (as shown by their anatomy) were present at the same time in widely separated parts of the world, it is highly probable that there was no barrier to their movement in one or both directions between the localities in the past.
- The most effective barrier to the movement of land animals (particularly mammals) was a sea between continents (as was caused by changing sea levels during the ice ages).
- The scattered distribution of modern species may be explained by the movement out of the area they originally occupied, or extinction in those regions between modern species.

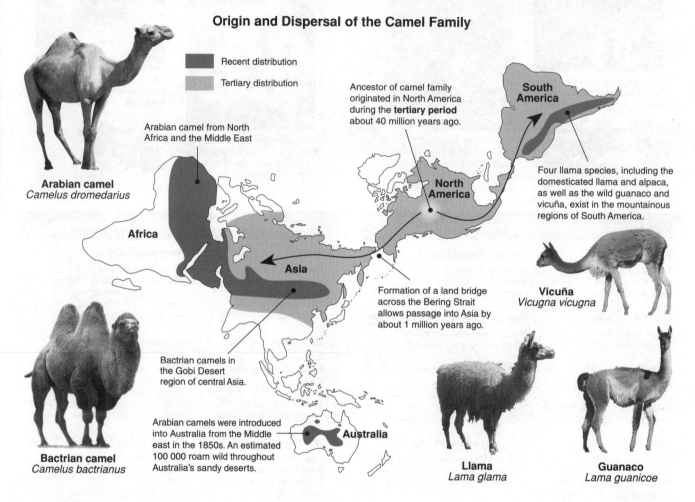

Origin and Dispersal of the Camel Family

Recent distribution
Tertiary distribution

Ancestor of camel family originated in North America during the **tertiary period** about 40 million years ago.

Arabian camel from North Africa and the Middle East

Arabian camel
Camelus dromedarius

Africa

Asia

South America

North America

Four llama species, including the domesticated llama and alpaca, as well as the wild guanaco and vicuña, exist in the mountainous regions of South America.

Formation of a land bridge across the Bering Strait allows passage into Asia by about 1 million years ago.

Vicuña
Vicugna vicugna

Bactrian camels in the Gobi Desert region of central Asia.

Bactrian camel
Camelus bactrianus

Arabian camels were introduced into Australia from the Middle east in the 1850s. An estimated 100 000 roam wild throughout Australia's sandy deserts.

Australia

Llama
Lama glama

Guanaco
Lama guanicoe

1. The early camel ancestors were able to move into the tropical regions of Central and South America. Explain why this did not happen in southern Asia and southern Africa:

_____

2. Arabian camels are found wild in the Australian outback. Explain how they got there and why they were absent during prehistoric times:

_____

3. The camel family originated in North America. Explain why there are no camels in North America now:

_____

4. Explain how early camels managed to get to Asia from North America: _____

_____

5. Describe the present distribution of the camel family and explain why it is scattered (discontinuous):

_____

_____

© Biozone International 2006

Adaptive Radiation in Mammals

Adaptive radiation is diversification (both structural and ecological) among the descendants of a single ancestral group to occupy different niches. Immediately following the sudden extinction of the dinosaurs, the mammals underwent an adaptive radiation. Most of the modern mammalian groups became established very early. The diagram below shows the divergence of the mammals into major orders; many occupying niches left vacant by the dinosaurs. The vertical extent of each grey shape shows the time span for which that particular mammal order has existed (note that the scale for the geological time scale in the diagram is not linear). Those that reach the top of the chart have survived to the present day. The width of a grey shape indicates how many species were in existence at any given time (narrow means there were few, wide means there were many). The dotted lines indicate possible links between the various mammal orders for which there is no direct fossil evidence.

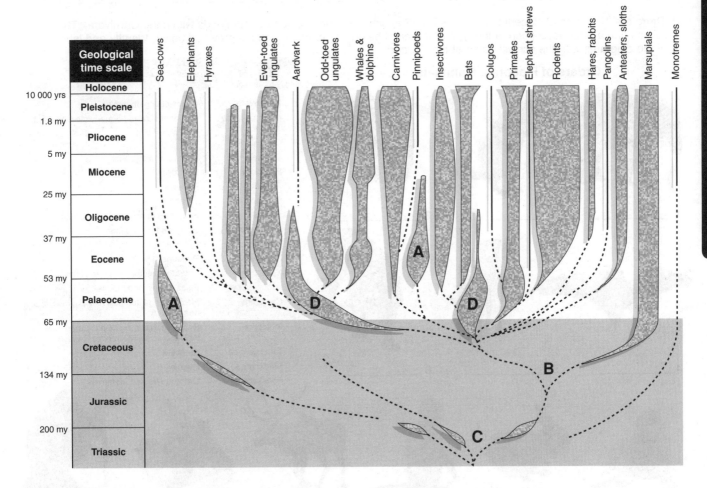

1. In general terms, discuss the **adaptive radiation** that occurred in mammals: _____

    _____

    _____

2. Name the term that you would use to describe the animal groups at point **C** (above): _____

3. Explain what occurred at point **B** (above): _____

    _____

4. Describe two things that the animal orders labelled **D** (above) have in common:

    (a) _____

    (b) _____

5. Identify the two orders that appear to have been most successful in terms of the number of species produced:

    _____

6. Explain what has happened to the mammal orders labelled **A** in the diagram above: _____

    _____

7. Identify the **epoch** during which there was the most adaptive radiation: _____

8. Describe two key features that distinguish mammals from other vertebrates:

    (a) _____ (b) _____

9. Describe the principal reproductive features distinguishing each of the major mammalian lines (sub-classes):

    (a) Monotremes: _____

    (b) Marsupials: _____

    (c) Placentals: _____

10. There are 18 orders of placental mammals (or 17 in schemes that include the pinnipeds within the Carnivora). Their names and a brief description of the type of mammal belonging to each group is provided below. Identify and label each of the diagrams with the correct name of their Order:

Orders of Placental Mammals

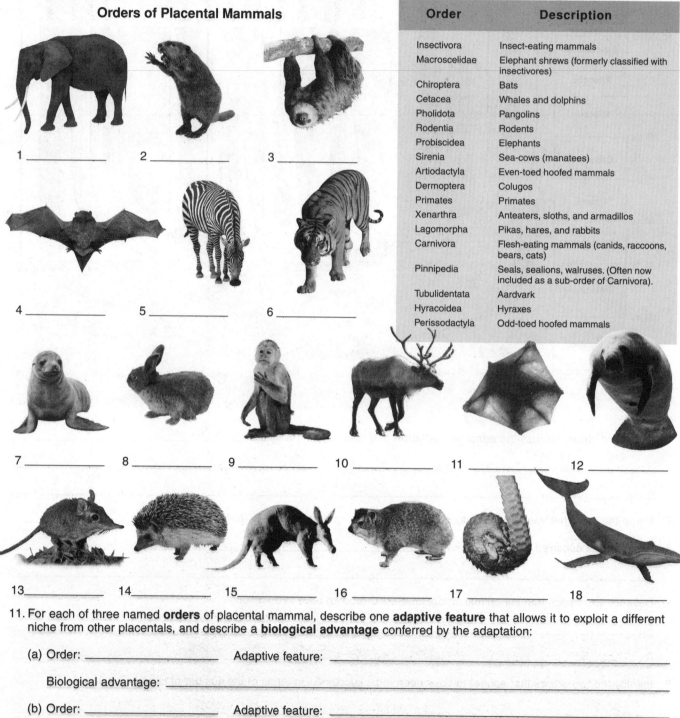

Order	Description
Insectivora	Insect-eating mammals
Macroscelidae	Elephant shrews (formerly classified with insectivores)
Chiroptera	Bats
Cetacea	Whales and dolphins
Pholidota	Pangolins
Rodentia	Rodents
Probiscidea	Elephants
Sirenia	Sea-cows (manatees)
Artiodactyla	Even-toed hoofed mammals
Dermoptera	Colugos
Primates	Primates
Xenarthra	Anteaters, sloths, and armadillos
Lagomorpha	Pikas, hares, and rabbits
Carnivora	Flesh-eating mammals (canids, raccoons, bears, cats)
Pinnipedia	Seals, sealions, walruses. (Often now included as a sub-order of Carnivora).
Tubulidentata	Aardvark
Hyracoidea	Hyraxes
Perissodactyla	Odd-toed hoofed mammals

1 _____ 2 _____ 3 _____

4 _____ 5 _____ 6 _____

7 _____ 8 _____ 9 _____ 10 _____ 11 _____ 12 _____

13 _____ 14 _____ 15 _____ 16 _____ 17 _____ 18 _____

11. For each of three named **orders** of placental mammal, describe one **adaptive feature** that allows it to exploit a different niche from other placentals, and describe a **biological advantage** conferred by the adaptation:

    (a) Order: _____ Adaptive feature: _____

    Biological advantage: _____

    (b) Order: _____ Adaptive feature: _____

    Biological advantage: _____

    (c) Order: _____ Adaptive feature: _____

    Biological advantage: _____

Adaptive Radiation in Ratites

The **ratites** evolved from a single common ancestor; they are a monophyletic group of birds that lost the power of flight very early on in their evolutionary development. Ratites possess two features distinguishing them from other birds: a flat breastbone (instead of the more usual keeled shape) and a primitive palate (roof to the mouth). Flightlessness in itself is not unique to this group. There are other examples of birds that have lost the power of flight, particularly on remote, predator-free islands. Fossil evidence indicates that the ancestors of ratites were flying birds living about 80 million years ago. These ancestors also had a primitive palate, but they possessed a keeled breastbone.

Elephantbird
Several species, extinct, Madagascar

Ostrich
Struthio camelus, Africa.

Emu
Dromaius novaehollandiae, Australia.

Cassowary
Three species, Australia & New Guinea.

Rhea
Two species, South America.

Kiwi
Three species, New Zealand.

Moa
Eleven species (Lambert *et al.* 2004*), all extinct, New Zealand.

The geographical distribution of modern day and extinct ratite species can be partially explained in terms of continental drift. The ancestral ratite population existed at a time when the southern continents of South America, Africa and Australia (together with their major offshore islands) were joined as a single land mass called Gondwana. As the continents moved apart as a result of plate tectonics, the early ratite populations were carried with them. Subsequent speciation on each continent and some of the islands produced the variety of forms shown here. The 50 species of tinamou (see chart below) from South America, are considered a sister group to the ratites even though they can fly, because they possess the archaic palate. This relationship is confirmed by DNA sequence tests. The diagram below shows a possible phylogenetic tree based upon comparisons of mitochondrial DNA sequences. This view has been supported by the extensive comparison of skeletons from the different ratite species.

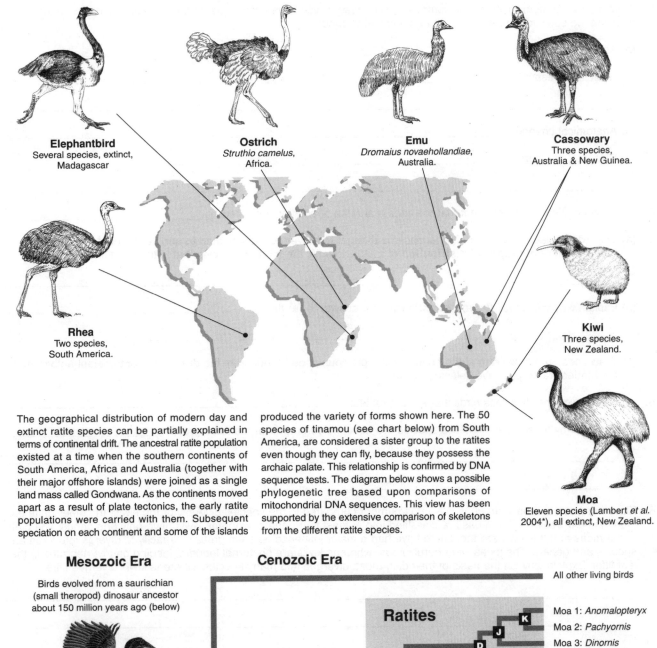

Mesozoic Era

Birds evolved from a saurischian (small theropod) dinosaur ancestor about 150 million years ago (below)

Ratites diverge from the line to the rest of the birds about 100 million years ago.

Cenozoic Era

Fossil evidence suggests that **ratite ancestors** possessed a keeled breastbone and an archaic palate (roof of mouth)

All other living birds
Moa 1: *Anomalopteryx*
Moa 2: *Pachyornis*
Moa 3: *Dinornis*
Moa 4: *Megalapteryx*
Little spotted kiwi
Great spotted kiwi
Brown kiwi
Emu
Cassowary
Ostrich
Rhea 1
Rhea 2
Tinamou (can fly)

Ratites

A Letters indicate common ancestors

* Lambert *et al.* 2004. "Ancient DNA solves sex mystery of moa." Australasian Science, 25(8), Sept. 2004, pp. 14-16.

Code: RDA 3

1. (a) Describe three physical features distinguishing all ratities from most other birds: _____
   _____

   (b) Identify the primitive feature shared by ratites and tinamou: _____
   _____

2. Describe two anatomical changes, common to all ratites, which have evolved as a result of flightlessness. For each, describe the selection pressures for the anatomical change:

   (a) Anatomical change: _____

   Selection pressure: _____
   _____

   (b) Anatomical change: _____

   Selection pressure: _____
   _____

3. Name the ancient supercontinent that the ancestral ratite population inhabited: _____

4. (a) The extinct elephantbird from Madagascar is thought to be very closely related to another modern ratite. Based purely on the **geographical distribution** of ratites, identify the modern species that is the most likely relative:
   _____

   (b) Explain why you chose the modern ratite in your answer to (a) above: _____
   _____

   (c) Draw lines on the diagram at the bottom of the previous page to represent the divergence of the elephantbird from the modern ratite you have selected above.

5. (a) Name two other flightless birds that are not ratites: _____

   (b) Explain why these other flightless species are not considered part of the ratite group: _____
   _____
   _____

6. Eleven species of moa is an unusually large number compared to the species diversity of the kiwis, the other ratite group found in New Zealand. The moas are classified into at least four genera, whereas kiwis have only one genus. The diets of the moas and the kiwis are thought to have had a major influence on each group's capacity to diverge into separate species and genera. The moas were herbivorous, whereas kiwis are nocturnal feeders, feeding on invertebrates in the leaf litter. Explain why, on the basis of their diet, moas diverged into many species, whereas kiwis diverged little:
   _____
   _____
   _____

7. The DNA evidence suggests that New Zealand had two separate invasions of ratites, an early invasion from the moas (before the breakup of Gondwana) followed by a second invasion of the ancestors of the kiwis. Describe a possible sequence of events that could account for this:
   _____
   _____
   _____
   _____

8. The common ancestors of divergent groups are labelled (A-L) on the diagram at the bottom of the previous page. State the **letter** identifying the **common ancestor** for:

   (a) The kiwis and the Australian ratites: _____ (b) The kiwis and the moas: _____

Origin of New Zealand Parrots

Recent **mitochondrial DNA** (mtDNA) studies at Victoria University confirm the existence in New Zealand of two distinct groups of parrots: kakapo-kaka-kea, and the various kakariki (five species). This research provides an excellent example of the use of **DNA analysis** to determine evolutionary relationships. The first group originated from an, as yet unknown, Australian ancestor about 100 million years ago (Mya). The formation of this proto-kaka/kakapo resulted-from the break-up of Gondwana, with New Zealand moving away in isolation from the Australian segment. The kakapo split from this lineage 60-80 million years ago and is today our most ancient parrot. Kaka split from the kea line some 3 Mya and an early member migrated to produce the now-extinct Norfolk Island kaka. About 400 000 years ago the North and South Island kaka differentiated. Mitochondrial DNA studies also show that the second group of New Zealand's parrots, the kakariki, are most likely derived from a New Caledonian parakeet ancestor, which in turn was derived at some time from an unknown Australian ancestor (possibly a proto-rosella). After the ancestral kakariki arrived in New Zealand, probably via Norfolk or Lord Howe Islands, it underwent speciation by migration-and-isolation (a process known as **vicariance**), and by ecological and behavioural divergence.

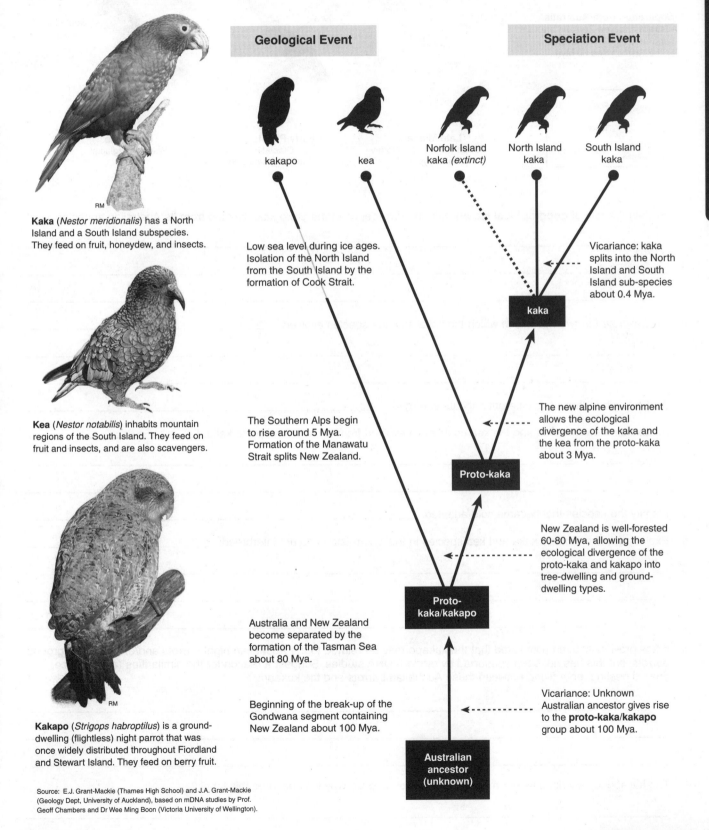

Geological Event

Speciation Event

kakapo   kea   Norfolk Island kaka *(extinct)*   North Island kaka   South Island kaka

Kaka (*Nestor meridionalis*) has a North Island and a South Island subspecies. They feed on fruit, honeydew, and insects.

Low sea level during ice ages. Isolation of the North Island from the South Island by the formation of Cook Strait.

Vicariance: kaka splits into the North Island and South Island sub-species about 0.4 Mya.

kaka

Kea (*Nestor notabilis*) inhabits mountain regions of the South Island. They feed on fruit and insects, and are also scavengers.

The Southern Alps begin to rise around 5 Mya. Formation of the Manawatu Strait splits New Zealand.

The new alpine environment allows the ecological divergence of the kaka and the kea from the proto-kaka about 3 Mya.

Proto-kaka

New Zealand is well-forested 60-80 Mya, allowing the ecological divergence of the proto-kaka and kakapo into tree-dwelling and ground-dwelling types.

Proto-kaka/kakapo

Kakapo (*Strigops habroptilus*) is a ground-dwelling (flightless) night parrot that was once widely distributed throughout Fiordland and Stewart Island. They feed on berry fruit.

Australia and New Zealand become separated by the formation of the Tasman Sea about 80 Mya.

Beginning of the break-up of the Gondwana segment containing New Zealand about 100 Mya.

Vicariance: Unknown Australian ancestor gives rise to the **proto-kaka/kakapo** group about 100 Mya.

Australian ancestor (unknown)

Source: E.J. Grant-Mackie (Thames High School) and J.A. Grant-Mackie (Geology Dept, University of Auckland), based on mDNA studies by Prof. Geoff Chambers and Dr Wee Ming Boon (Victoria University of Wellington).

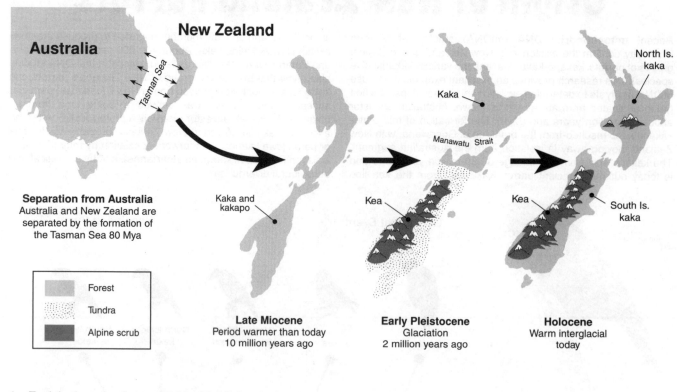

New Zealand

Australia

Tasman Sea

Separation from Australia
Australia and New Zealand are
separated by the formation of
the Tasman Sea 80 Mya

	Forest
	Tundra
	Alpine scrub

Kaka and
kakapo

Late Miocene
Period warmer than today
10 million years ago

Kaka

Manawatu Strait

Kea

Early Pleistocene
Glaciation
2 million years ago

North Is.
kaka

Kea

South Is.
kaka

Holocene
Warm interglacial
today

1. Explain the role of **geographical isolation** in the divergence of the proto-kaka/kakapo from its Australian ancestor:

_____

_____

_____

2. (a) Describe the two habitats in which the kaka and kea species evolved: _____

_____

_____

   (b) State when these two different habitats emerged: _____

   (c) Explain how the kaka and the kea could have evolved from an ancestral proto-kaka: _____

_____

_____

3. Identify the species that became cold-adapted: _____

4. Explain why the modern kaka and kea species in the South Island do not interbreed: _____

_____

_____

_____

5. It has previously been postulated that the kakapo may be related to the Australian night parrots and/or Australian ground
   parrots, but this has not been supported by modern DNA studies. Suggest a reason for the similarities (appearance,
   ground nesting, poor flight) between these Australian parrots and the kakapo:

_____

_____

_____

6. The kakapo is described as our most ancient parrot. Explain why they deserve this label: _____

_____

Adaptive Radiation in Wrens

The New Zealand wrens have been isolated from their probable ancestral stock in Australia for more than 60 million years. The period of isolation has been sufficient for an endemic family (Family: Acanthisittidae) to have developed (see note, below right). This family includes the rifleman, the rock wren, and the bush wren, as well as a number of extinct species. Characteristics include small size, flightless or with poor flying ability, and insectivorous. Until recent times, their distribution was throughout New Zealand, but now it is much more restricted. One fossil deposit in Nelson has yielded six species of wren found together. Comparative anatomy of leg muscles and recent DNA hybridisation studies suggest that these birds are an ancient, isolated group with a common ancestry. The ancestral wren was almost certainly insectivorous since all the descendants are. There may well have been several related species, of which the ancestral wren was one, living in various habitats throughout

early New Zealand. The existing wren family dates from a time of extensive adaptive radiation some 20 Mya. This adaptive radiation was a consequence of the break up of the New Zealand land mass during the Oligocene, (see following page), had a profound effect on the evolution of the wrens. Earlier species, as yet undiscovered, may also have existed at this time. Although all insectivorous, the living and extinct wrens have exploited different habitats and feeding niches (see below). The fossils of extinct species can illustrate past land connections. Fossils of the Stephens Is. wren (extinct) have been found in both the North and the South Islands, showing that there was a land connection between these islands in the past. The two species of stout-legged wren may have undergone allopatric speciation 3-5 million years ago when changing sea levels separated the North and South Islands. Later similar events could also explain the evolution of separate sub-species of bush wrens and rifleman.

Curved beak wren
Spent time scurrying up and down tree trucks, probing in crevices for grubs with its curved beak.

Stout legged wren
Both species were ground dwelling, searching the ground for insects.

Stephens Is. wren
Lived and foraged in grass and underbrush (ecological equivalent of a field mouse).

Bush wren (recently extinct)
Lived and foraged in the bush searching for insects on the ground and in the air.

Rock wren
Lives in subalpine areas, feeding among rocks and tussock, and surviving winter under the snow layer.

Rifleman
Lives and feeds in the bush, picking insects from the branches and trunks of trees.

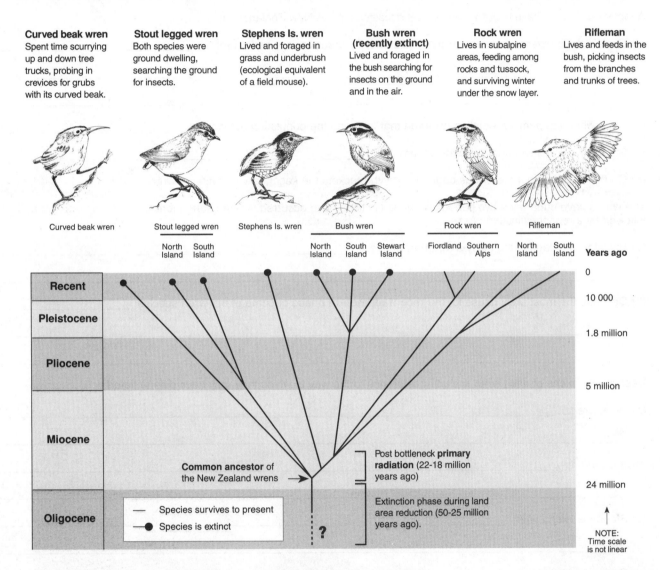

Curved beak wren

Stout legged wren
North Island South Island

Stephens Is. wren

Bush wren
North Island South Island Stewart Island

Rock wren
Fiordland Southern Alps

Rifleman
North Island South Island

Years ago

Recent	0
	10 000
Pleistocene	
	1.8 million
Pliocene	
	5 million
Miocene	
	24 million
Oligocene	

Common ancestor of the New Zealand wrens →

Post bottleneck **primary radiation** (22-18 million years ago)

Extinction phase during land area reduction (50-25 million years ago).

— Species survives to present
● Species is extinct

?

NOTE: Time scale is not linear

New Zealand Wren Classification

Rifleman, North Island (NI)	Acanthisitta chloris granti	Can fly	
Rifleman, South Island (SI)	Acanthisitta chloris chloris	Can fly	
Rock wren (Southern Alps)	Xenicus gilviventris	Can fly	
Rock wren (Fiordland)	Xenicus gilviventris riney	Can fly	
Bush wren (NI)	Xenicus longipes stokesi	Could fly	Presumed extinct
Bush wren (SI)	Xenicus longipes longipes	Could fly	Presumed extinct
Stead's bush wren (Stewart Is.)	Xenicus longipes variabilis	Could fly	Extinct
Stephens Is. wren	Traversia lyalli	Flightless	Extinct
Stout-legged wren (SI)	Pachyplichas yaldwyni	Flightless	Extinct
Stout-legged wren (NI)	Pachyplichas jagmi	Flightless	Extinct
Curved beak wren	Dendroscansor decurvirostris	Flightless	Extinct

NOTE: On the diagram, the timescale for the emergence of new sub-species, species, and genera. As a general rule among birds, new sub-species emerge after 0.5-2 million years of separation, new species after 2-10 million years, new genera after 10-20 million years and new orders with 60-80 million years of separation.

Many thanks to **Ewan Grant-Mackie**, Thames High School, and **Prof. J.A. Grant-Mackie**, Geology Dept, Auckland University, who supplied the information for this exercise.

The New Zealand Land Mass

Between 25-30 million years ago, during the Oligocene period, New Zealand was almost completely submerged, and existed only as a chain of small islands, with a land mass only 18% of what it is today. This was the result of rising sea levels and land subsidence over a period of 5 million years.

The New Zealand Wrens

The reduction in size and break-up of the single land mass had a profound effect on the evolution of the wrens, greatly reducing their range of habitats and causing selective extinctions:

- *The reduction in species diversity*
When the New Zealand land mass was inundated during the Oligocene, some animals would have retreated to islands of high ground, but would have perished as these became submerged. As a result, a great many species were lost (species diversity declined).

- *The reduction in genetic diversity*
Not only was species diversity lost during this period, the genetic diversity of remaining populations would have been severely depleted, with few individuals in a species surviving to pass on their genes. This situation, referred to as a **population bottleneck**, occurs when a very small sample of the total species gene pool manages to survive.

When the sea levels dropped again, the survivors moved into new areas to occupy newly available niches.

New Zealand During the Oligocene
25 – 30 million years ago

New Zealand shoreline

▨ Oligocene

☐ Present day

1. Adaptive radiations have occurred on several occasions in the New Zealand wrens:

   (a) Explain the difference between **primary radiation** and **secondary radiation**: _____

   _____

   _____

   (b) State when the primary radiation occurred that following the population bottleneck: _____

   (c) State when the secondary radiation occurred: _____

   (d) On the diagram on the previous page, mark with brackets the secondary radiation of wrens.

2. The wrens have undergone two periods where extinctions have occurred; an early one more than 25 million years ago, followed by a recent extinction phase.

   (a) Explain what caused the **early extinction** phase: _____

   _____

   (b) Describe the most likely cause for the **recent extinction** phase of some of the wren species: _____

   _____

   _____

3. Describe the niche of each wren, including reference to the way in which they may have differentiated:

   (a) Rock wren: _____

   _____

   (b) Bush wren: _____

   _____

   (c) Stephens Island wren: _____

   _____

   (d) Curved beak wren: _____

   _____

   (e) Rifleman: _____

   _____

4. Explain how geological events in New Zealand affected the radiation of the wrens: _____

   _____

   _____

Evolution in Springtails

Researchers wanted to investigate the genetic relatedness of springtails in a Dry Valley in Antarctica. These tiny arthropods have a limited capacity to move between locations, apart from sudden movements when they may be swept along by glacial runoff during the summer melt. Results of a mtDNA study show two distinct genetic 'types' of springtail in Taylor Valley (see map, black and white squares on the following page). The two types have different DNA bases at a number of positions in a mitochondrial gene. They also coexist in an area of **sympatry** in the middle of Taylor Valley. The results of the research are summarised below. It shows an order of separation based on the genetic differences between the two types (TV1-14) compared with other populations of the same species (from Cape Evans, Cape Royds, and Beaufort Is). One other Antarctic species of springtail (*Biscoia sudpolaris*) is included on the diagram as an 'outgroup' (reference point). The genetic difference between populations is indicated by the distance to the 'branching point'. Groups that branch apart early in the tree are more genetically different than groups that branch later.

Patterns of Evolution

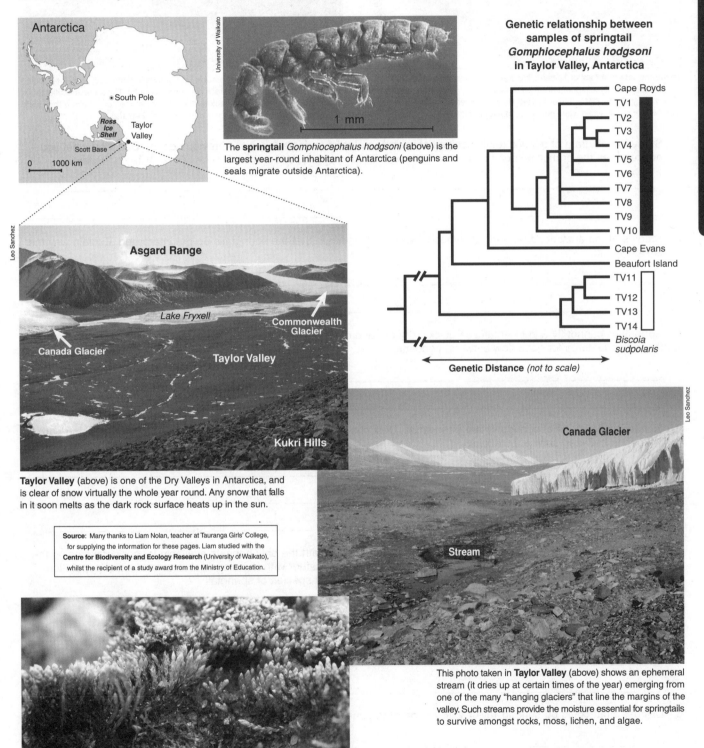

The **springtail** *Gomphiocephalus hodgsoni* (above) is the largest year-round inhabitant of Antarctica (penguins and seals migrate outside Antarctica).

Taylor Valley (above) is one of the Dry Valleys in Antarctica, and is clear of snow virtually the whole year round. Any snow that falls in it soon melts as the dark rock surface heats up in the sun.

Source: Many thanks to Liam Nolan, teacher at Tauranga Girls' College, for supplying the information for these pages. Liam studied with the **Centre for Biodiversity and Ecology Research** (University of Waikato), whilst the recipient of a study award from the Ministry of Education.

This photo taken in **Taylor Valley** (above) shows an ephemeral stream (it dries up at certain times of the year) emerging from one of the many "hanging glaciers" that line the margins of the valley. Such streams provide the moisture essential for springtails to survive amongst rocks, moss, lichen, and algae.

Mosses (left) are the tallest plants in Antarctica. They provide ideal habitats for springtails. Although springtails have antifreeze (glycerol) in their blood, they are still vulnerable to freezing. Antarctic springtails do not possess the proteins that some Antarctic fish have to help them avoid freezing.

Code: RA 3

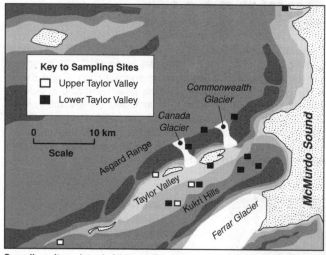

Sampling sites: A total of 14 sampling sites was used to build up a picture of the genetic diversity of springtails in an area of Taylor Valley. They were named TV1 through TV14 (TV = Taylor Valley). Black squares represent one genetic 'type' of springtail, while white squares represent another.

Liam Nolan uses a pooter to suck up springtails disturbed from their hiding places between the flakes of rock covering a boulder. The samples collected in Taylor Valley, were brought back to the "lab tent" at the camp, where they were preserved and prepared for their return to New Zealand.

1. Study the diagram of genetic relationships between samples of springtails (on the previous page). Describe what you notice about the branching point of the populations from the upper (TV11-14) and lower (TV1-10) Taylor Valley:

   _____

   _____

2. Studies of the enzymes from the two 'types' of springtails indicate that the springtails do not interbreed. Explain why this is significant:

   _____

   _____

3. Springtails cannot fly and in Antarctica quickly dry out and die if they are blown by the wind. Discuss the significance of these two features for gene flow between populations:

   _____

   _____

4. Isolated populations of springtails are often small. Describe the mechanism that could be important in increasing genetic difference between such populations:

   _____

   _____

   _____

5. Taylor Valley was once (thousands of years ago) covered in ice, with the only habitats available for springtails being the mountain tops lining both sides of the valley. Explain how this, together with low dispersal rates and small population size, could result in the formation of two species from one original species of springtail:

   _____

   _____

   _____

   _____

6. Discuss what could be done in Taylor Valley in order to conserve the biodiversity of springtails:

   _____

   _____

   _____

   _____

Ancient New Zealand Landscapes

Patterns of Evolution

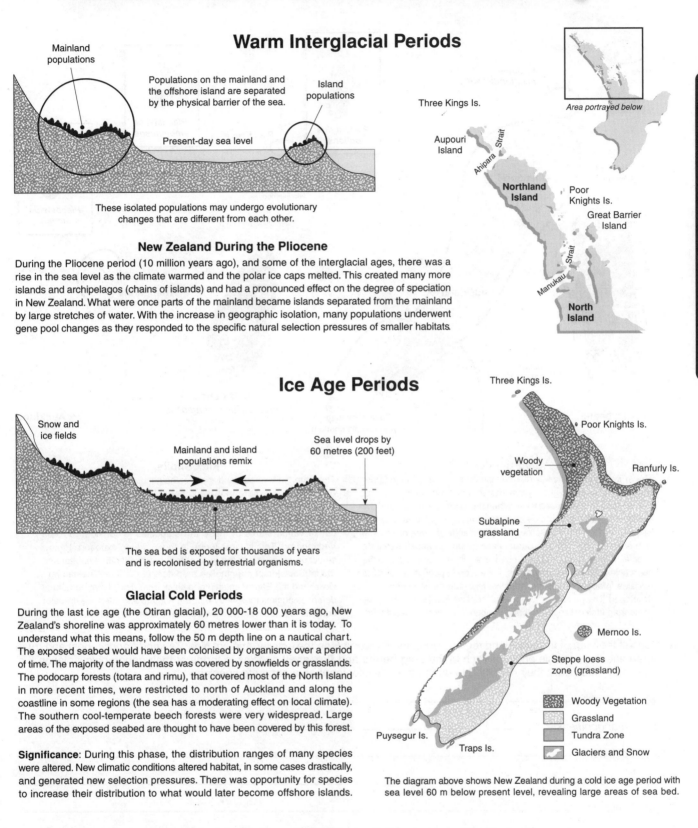

Warm Interglacial Periods

Mainland populations

Populations on the mainland and the offshore island are separated by the physical barrier of the sea.

Island populations

Present-day sea level

These isolated populations may undergo evolutionary changes that are different from each other.

Three Kings Is.

Area portrayed below

Aupouri Island

Ahipara Strait

Northland Island

Poor Knights Is.

Great Barrier Island

Manukau Strait

North Island

New Zealand During the Pliocene

During the Pliocene period (10 million years ago), and some of the interglacial ages, there was a rise in the sea level as the climate warmed and the polar ice caps melted. This created many more islands and archipelagos (chains of islands) and had a pronounced effect on the degree of speciation in New Zealand. What were once parts of the mainland became islands separated from the mainland by large stretches of water. With the increase in geographic isolation, many populations underwent gene pool changes as they responded to the specific natural selection pressures of smaller habitats.

Ice Age Periods

Snow and ice fields

Mainland and island populations remix

Sea level drops by 60 metres (200 feet)

The sea bed is exposed for thousands of years and is recolonised by terrestrial organisms.

Three Kings Is.

Poor Knights Is.

Woody vegetation

Ranfurly Is.

Subalpine grassland

Mernoo Is.

Steppe loess zone (grassland)

Glacial Cold Periods

During the last ice age (the Otiran glacial), 20 000-18 000 years ago, New Zealand's shoreline was approximately 60 metres lower than it is today. To understand what this means, follow the 50 m depth line on a nautical chart. The exposed seabed would have been colonised by organisms over a period of time. The majority of the landmass was covered by snowfields or grasslands. The podocarp forests (totara and rimu), that covered most of the North Island in more recent times, were restricted to north of Auckland and along the coastline in some regions (the sea has a moderating effect on local climate). The southern cool-temperate beech forests were very widespread. Large areas of the exposed seabed are thought to have been covered by this forest.

Significance: During this phase, the distribution ranges of many species were altered. New climatic conditions altered habitat, in some cases drastically, and generated new selection pressures. There was opportunity for species to increase their distribution to what would later become offshore islands.

Woody Vegetation

Grassland

Tundra Zone

Glaciers and Snow

Puysegur Is.

Traps Is.

The diagram above shows New Zealand during a cold ice age period with sea level 60 m below present level, revealing large areas of sea bed.

1. Explain how the warmer interglacial periods contributed to speciation events in New Zealand: _____

_____

_____

2. Explain how the ice ages contributed to speciation events in New Zealand: _____

_____

_____

Evolution in NZ Invertebrates

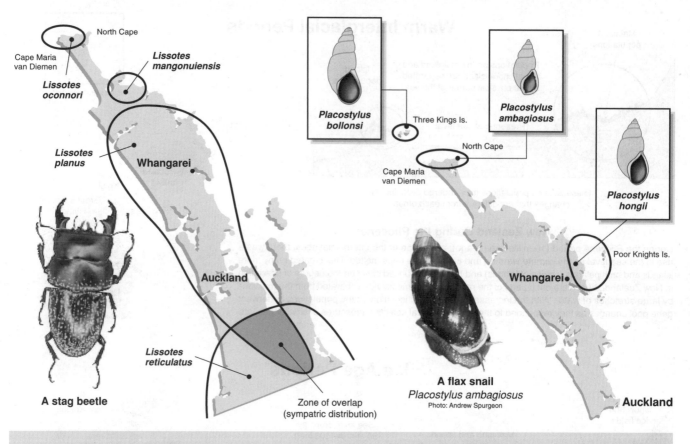

North Cape

Cape Maria van Diemen

Lissotes oconnori

Lissotes mangonuiensis

Lissotes planus

Whangarei

Lissotes reticulatus

A stag beetle

Auckland

Zone of overlap (sympatric distribution)

Placostylus bollonsi

Three Kings Is.

Placostylus ambagiosus

North Cape

Cape Maria van Diemen

Placostylus hongii

Poor Knights Is.

Whangarei

A flax snail
Placostylus ambagiosus
Photo: Andrew Spurgeon

Auckland

Stag Beetles

The four species of closely related stag beetles found in the upper North Island are thought to have shared a common ancestor prior to the Pliocene rise in sea level. When the climate warmed and the sea level rose, a chain of islands (an archipelago) was created, isolating parts of the population. Over thousands of years of isolation with different selection pressures, each group developed separate species status. Since the fall in sea level, these populations have been able to remix in many cases, but the gene pool of the species remains intact because reproductive isolating mechanisms have developed. The point of origin for each species can be estimated by comparing their present distribution with that of Pliocene Northland.

Flax Snails

The large, herbivorous land snails of the genus *Placostylus* are found on islands elsewhere in the southern Pacific Ocean, but in New Zealand they are restricted to localities in the North Auckland Peninsula and off-lying island conservation reserves. The present conservation status of these three species is considered to be endangered. This is largely due to their sudden exposure to new selection pressures in the form of reduced or modified habitat and the introduction of mammalian predators (e.g. the polynesian rat - kiore, and the European ship and black rats). They have no natural defences against these very efficient hunters and intensive predator controls are necessary if their populations are to recover.

1. Consult the map of the Northland region during the Pliocene (see the activity *Ancient New Zealand Landscapes*) and try to determine the possible origin of each of the **stag beetle** and **flax snail** species in terms of their geographic isolation (i.e. what land mass were they restricted to during their development as a separate species).

   (a) *L. oconnori*: _____

   (b) *L. mangonuiensis*: _____

   (c) *L. planus*: _____

   (d) *L. reticulatus*: _____

   (e) *P. bollonsi*: _____

   (f) *P. ambagiosus*: _____

   (g) *P. hongii*: _____

2. (a) Identify the two species of stag beetle that are sympatric in their distribution: _____

   _____

   (b) Describe the likely event that allowed these two species to occupy the same region south of Auckland:

   _____

Evolution in Hebe

A large part of New Zealand's angiosperm flora is made up of groups that have undergone spectacular adaptive radiations throughout the country. The *Hebe* group of plants consists of five genera, comprising some 150 species, and are considered to be **monophyletic** (i.e. they share a single common ancestor). The majority of the species in the group are in a single genus, *Hebe*, with about 100 species currently recognised. About 20 of these remain undescribed as a result of their uncertain taxonomic status. Most of these undescribed species have polyploid chromosome numbers, resulting in **instant speciation**. This suggests that the *Hebe* group is still evolving rapidly. Recent

molecular studies suggest that hebes arrived in New Zealand less than 5 million years ago from an ancestor in Gondwana, or migration from South-east Asia via New Guinea. These same studies also suggest a surprising number of movements out of New Zealand back to Australia. The pattern of current *Hebe* distribution in the region of Mt. Taranaki (Mt. Egmont) illustrates an interesting phenomenon; some of the *Hebe* species found at high altitude may have first colonised the once high slopes of Kaitake and Pouakai from mountains elsewhere in the North Island. These remnants of extinct volcanoes once had summits considerably higher than their current eroded states suggest.

Profile of Mount Taranaki

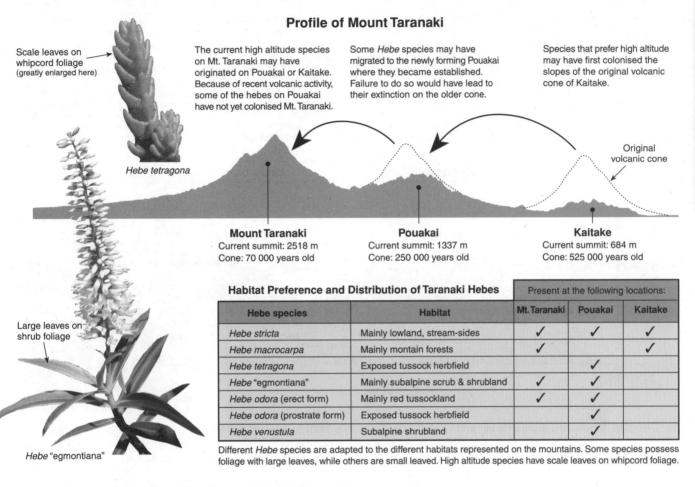

Scale leaves on whipcord foliage (greatly enlarged here)

Hebe tetragona

Large leaves on shrub foliage

Hebe "egmontiana"

The current high altitude species on Mt. Taranaki may have originated on Pouakai or Kaitake. Because of recent volcanic activity, some of the hebes on Pouakai have not yet colonised Mt. Taranaki.

Some *Hebe* species may have migrated to the newly forming Pouakai where they became established. Failure to do so would have lead to their extinction on the older cone.

Species that prefer high altitude may have first colonised the slopes of the original volcanic cone of Kaitake.

Original volcanic cone

Mount Taranaki
Current summit: 2518 m
Cone: 70 000 years old

Pouakai
Current summit: 1337 m
Cone: 250 000 years old

Kaitake
Current summit: 684 m
Cone: 525 000 years old

Habitat Preference and Distribution of Taranaki Hebes

Hebe species	Habitat	Mt. Taranaki	Pouakai	Kaitake
Hebe stricta	Mainly lowland, stream-sides	✓	✓	✓
Hebe macrocarpa	Mainly montain forests	✓		✓
Hebe tetragona	Exposed tussock herbfield		✓	
Hebe "egmontiana"	Mainly subalpine scrub & shrubland	✓	✓	
Hebe odora (erect form)	Mainly red tussockland	✓	✓	
Hebe odora (prostrate form)	Exposed tussock herbfield		✓	
Hebe venustula	Subalpine shrubland		✓	

Different *Hebe* species are adapted to the different habitats represented on the mountains. Some species possess foliage with large leaves, while others are small leaved. High altitude species have scale leaves on whipcord foliage.

1. Explain what is meant by the term **monophyletic**: _____

_____

2. Describe the process that resulted in **instant speciation** for a significant number of the undescribed *Hebe* species:

_____

3. Explain how high altitude species may have first colonised the slopes of Kaitake, despite its low (684 m) summit:

_____

_____

4. Describe two features of the hebes in Taranaki that support the notion that the group has undergone adaptive radiation:

_____

_____

5. Although Pouakai has a lower altitude than Mt. Taranaki, it has a greater number of *Hebe* species in its slopes. Explain this apparent anomaly:

_____

Code: R 2

Extinction

Extinction is an important process in evolution as it provides opportunities, in the form of vacant niches, for the development of new species. Most species that have ever lived are now extinct. The species alive today make up only a fraction of the total list of species that have lived on Earth throughout its history. Extinction is a natural process in the life cycle of a species. Background extinction is the steady rate of species turnover in a taxonomic group (a group of related species). The duration of a species is thought to range from as little as 1 million years for complex larger organisms, to as long as 10-20 million years for simpler organisms. Superimposed on this constant background extinction are catastrophic events that wipe out vast numbers of species in relatively brief periods of time in geological terms. The diagram below shows how the number of species has varied over the history of life on Earth. The number of species is indicated on the graph by families; a taxonomic group comprising many genera and species. There have been five major extinction events and two of these have been intensively studied by palaeontologists.

Major Mass Extinctions

The Permian extinction
(225 million years ago)

This was the most devastating mass extinction of all. Nearly all life on Earth perished, with 90% of marine species and probably many terrestrial ones also, disappearing from the fossil record. This extinction event marks the **Palaeozoic-Mesozoic** boundary.

The Cretaceous extinction
(65 million years ago)

This extinction event marks the boundary between the Mesozoic and Cenozoic eras. More than half the marine species and many families of terrestrial plants and animals became extinct, including nearly all the dinosaur species (the birds are now known to be direct descendants of the dinosaurs).

Megafaunal extinction
(10 000 years ago)

This mass extinction occurred when many giant species of mammal died out. This is known as the Pleistocene overkill because their disappearance was probably hastened by the hunting activities of prehistoric humans. Many large marsupials in Australia and placental species elsewhere became extinct.

The sixth extinction
(now)

The current mass extinction is largely due to human destruction of habitats (e.g. coral reefs, tropical forests) and pollution. It is considered far more serious and damaging than some earlier mass extinctions because of the speed at which it is occurring. The increasing human impact is making biosphere recovery difficult.

Palaeozoic — Mesozoic — Cenozoic

Numbers of families: 700, 600, 500, 400, 300, 200, 100

Ordovician — Devonian — Permian — Triassic — Cretaceous

600 — 500 — 400 — 300 — 200 — 100 — 0

Millions of years ago

1. Describe the main features (scale and type of organisms killed off) of each of the following major extinction events:

   (a) Permian extinction: _____

   _____

   _____

   (b) Cretaceous extinction: _____

   _____

   _____

   _____

   (c) Megafaunal extinction: _____

   _____

   _____

2. Explain how human activity has contributed to the most recent mass extinction: _____

   _____

3. In general terms, describe the effect that past mass extinctions had on the way the surviving species **further evolved**:

   _____

   _____

© Biozone International 2006

Causes of Mass Extinctions

Over the last 540 million years, marine life has experienced about 24 bouts of mass extinction: five of these were major and some 19 were minor (judged by the percentage of genera that became extinct). Many of these extinction peaks coincided with known comet/asteroid impact events, strongly implying that they may have been the cause. A general deterioration of the environment, caused by climatic change or some major cosmic, geological, or biological event, is deemed the principal cause of extinctions. The ability of the biosphere to recover from such crises is evident by the fact that life continues to exist today.

Large asteroid/comet impacts   NASA   Large solar flares   Volcanism

Possible causes of mass extinction	Effects on the extinction rate	Examples of extinction events and their likely causes
Impacts by large asteroid/comet or 'showers': Shock waves, heat-waves, wildfires, impact 'winters' caused by global dust clouds, super-acid rain, toxic oceans, superwaves and superfloods from an oceanic impact.	Global extinction of much of the planet's biodiversity. Smaller comet showers could cause stepwise, regional extinctions.	**Late Pleistocene** (15 000-10 000 ya) *Extinction of:* Many large mammals and large flightless birds. *Probable cause:* Warming of the global climate after the last ice age plus predation (hunting) by humans.
Supernovae radiation: Direct exposure to X-rays and cosmic rays. Ozone depletion and subsequent exposure to excessive UV radiation from the sun.	Causes mutations and kills organisms. Selective mass extinctions, particularly of animals (but not plants) exposed to the atmosphere, as well as shallow-water aquatic forms.	**Late Cretaceous** (65 Mya) *Extinction of:* Dinosaurs, plesiosaurs, icthyosaurs, mosasaurs, pterosaurs, ammonites and belemnites (squid-like animals), and many other groups.
Large solar flares: Exposure to large doses of X-rays, and UV radiation. Ozone depletion.	Mass extinctions.	*Probable cause:* An asteroid impact (probably Yucatan peninsula) produced catastrophic environmental disturbance.
Geomagnetic reversals: Increased flux of cosmic rays.	Mass extinctions.	**Late Permian** (250 Mya) *Extinction of:* 90% of marine species and 70% of land species. Coral reefs, trilobites, some amphibians, and mammal-like reptiles, were eliminated (the Great Dying).
Continental drift: Climatic changes, such as glaciations and droughts, occur when continents move towards or away from the poles.	Global cooling due to changes in the pattern of oceans currents caused by shifting land masses. Extinctions as species find themselves in inhospitable climates.	*Probable cause:* An asteroid impact, followed by furious volcanic activity, a rapid heating of the atmosphere, and depletion of life-giving oxygen from the oceans.
Intense volcanism: Cold conditions caused by volcanic dust reducing solar input. Volcanic gases causing acid rain and reduced alkalinity of oceans. Toxic trace elements.	Stepwise mass extinctions.	**Late Devonian** (360 Mya) *Extinction of:* Many corals, bivalves, fish, sponges (21% of all marine families). Collapse of tropical reef communities.
Sea level change: Loss of habitat.	Mass extinctions of susceptible species (e.g. marine reptiles, coral reefs, coastal species).	*Probable cause:* Global cooling associated with (or causing?) widespread oxygen deficiency of shallow seas.
Arctic spill over: Release of cold fresh water or brackish water from an isolated Arctic Ocean. Ocean temperature falls 10°C, resulting in atmospheric cooling, drought.	Mass extinctions in marine ecosystems. Change of vegetation with drastic effect on large reptiles.	Source: *Evolution: A Biological and Palaeontological Approach*, Skelton, P. (ed.), Addison-Wesley (1993)
Anoxia: Shortage of oxygen.	Mass extinctions in the oceans.	
Spread of disease/predators: Direct effects due to changing geographic distribution.	Mass extinctions.	

Source: *Environmental Change – The Evolving Ecosphere* (1997), by R.J. Huggett

1. Describe how each of the following events might have caused mass extinctions in the past:

(a) Large asteroid/comet impact: _____

_____

(b) Continental drift: _____

_____

(c) Volcanism: _____

_____

2. Explain how the arrival of a new plant or animal species onto a continent may cause the demise of other species there:

_____

Code: A 3

Index